Sketching *for Engineering Design Visualization*

Title: Subtitle, Edition: Sketching for Engineering Design Visualization
Author(s): Jon M. Duff, William A. Ross

Vice President, Career and Professional Editorial: Dave Garza
Director of Learning Solutions: Sandy Clark
Acquisitions Editor: James Devoe
Managing Editor: Larry Main
Product Manager: John Fisher
Editorial Assistant: Thomas Best
Vice President, Career and Professional Marketing: Jennifer McAvey
Marketing Director: Deborah S. Yarnell
Marketing Manager: Jimmy Stephens
Marketing Coordinator: Mark Pierro
Production Director: Wendy Troeger
Production Manager: Mark Bernard
Content Product Manager: David Plagenza
Art Director: Bethany Casey
Technology Project Manager: Christopher Catalina
Production Technology Analyst: Thomas Stover
Cover Design: Bethany Casey
Interior Design: The WestHighland Press

Library of Congress Control Number: 2008931221

ISBN-13: 978-143545362X
ISBN-10: 143545362X

Delmar
5 Maxwell Drive
Clifton Park, NY 12065-2919
USA

Cengage Learning is a leading provider of customized learning solutions with office locations around the globe, including Singapore, the United Kingdom, Australia, Mexico, Brazil and Japan. Locate your local office at: **international.cengage.com/region**

Cengage Learning products are represented in Canada by Nelson Education, Ltd.

For your lifelong learning solutions, visit **delmar.cengage.com**
Visit our corporate website at **cengage.com**.

Printed in United States of America

2 3 4 5 6 7 10 09

Sketching *for Engineering Design Visualization*

Jon M. Duff
Arizona State University
William A. Ross
Purdue University

Contents

Preface

This sketching text-workbook is coordinated with *Visualization, Modeling, and Graphics for Engineers,* a landmark work by Dennis Lieu and Sheryl Sorby. Still, you will find the chapters in these pages follow their own order, a syllabus reflecting how Professors Duff and Ross have learned to teach over many decades and with thousands of engineering and technology students. This book is based on the concept that modern graphics may be started freehand and subsequently formalized by using CAD. In this book you will find a concise overview of basic CAD modeling techniques and a full complement of engineering drawing and modeling exercises through sketching. Chapters cover model visualization through orthographic multiviews, missing view problems, isometric pictorials, auxiliary views, section views, assemblies, dimensioning practices, and sketching for fabrication. Each chapter contains a brief overview followed by numerous sketching problems intended to serve as pre-CAD visualization exercises. If you desire more in-depth information, peruse the appropriate chapters in *Visualization, Modeling, and Graphics for Engineers.*

Sketching Exercises in Support of CAD Modeling

This text makes an ideal lab manual for a number of engineering design graphics texts. Because of the generic nature of the problems in this book, it may be used in conjunction with any 2D or 3D CAD software textbook. Although the emphasis here is on 3D modeling and visualization, the inclusion of traditional engineering drawing practices such as "orthographic reading" and missing view problems make it highly useful with 2D engineering graphics courses as well.

How to Use this Workbook

Two author philosophies heavily influenced the structure of this text-workbook and accepting them is important to you, the engineering and technical graphics student.

- Modeling is an introduction to sketching because you should view sketching as a modeling exercise—albeit limited by pencil and paper.

- Isometric sketching is necessary before orthographic reading because pictorial sketches are a major tool for visualizing geometry.

This book begins with an introduction to visualization—in this case, through rotations and mental cutting plane exercises. Terminology is presented and standard techniques used in 3D model building are introduced. All subsequent chapters then use these techniques and apply them to the standard practices and conventions used in engineering design graphics and technical drawing. The book is written so the reader will be able to visualize, sketch, and solve modeling geometry problems either prior to or during a work session with 3D CAD software. You will find the relationship between visualization, preliminary pencil sketching, and the modeling-drawing strengthened. In order to reinforce the value and appearance of planning with sketching, all illustrations, figures, examples, and exercises in this book are presented in a sketched format. The typical arrangement of material in each chapter includes:

- A brief overview and introduction to the chapter topic. This is often titled "General Conditions."

- Practical example from business and industry showing how the sketching topic is related to CAD.

Finally, sketching exercises are presented for practice and skill development. Because this is a supplementary text or lab manual, the bulk of each chapter is devoted to practical exercises progressing in difficulty from easy or moderate to complex.

Acknowledgements

We wish to thank the many individuals who contributed, over a period of nearly two decades, to the success of the first edition of this book. Special recognition goes to Dr. Ted Conway from the University of Akron, Dr. Larry Genalo of Iowa State University, Dr. Steve Howell at Northern Arizona University, Professor Menardi at Tufts University, Dr. Paul Morrison at the British Columbia Institute of Technology, Dr. Stephen Reed from Weber State University, Dr. Thomas Singer at Sinclair Community College, and Dr. Kelly Waldal at the University of Regina for their work on the first edition. The authors thank Dr. Dennis Lieu of the University of California-Berkeley and Dr. Sheryl Sorby of Michigan Technological University for bringing the authors' sketching experience into *Visualization, Modeling, and Graphics for Engineers.* Any mistakes in the printed version are not theirs, but solely the authors' responsibility.

Introduction

To Teachers

To assist your students, all exercises are placed on grids, scaled, and include world and user coordinate system icons. Once the student has visualized and sketched a solution, the completed sketches can be used to directly construct accurate three-dimensional (3D) models on all major CAD software programs. Throughout this text-workbook, 3D Cartesian coordinate systems and user-defined construction planes (UCS) are shown in conjunction with grids or dimensions for all sketching exercises. The purpose of identifying X, Y, and Z coordinate systems and construction planes on sketching exercises is to enable students to construct 3D computer models directly from sketched solutions. There is no standard icon used in CAD software to designate world and user coordinate systems. Because of its widespread use, and for consistency, the graphical representation of coordinate systems and gridded reference planes in this book is based on the familiar UCS icon utilized in AutoCAD software. The UCS icon, familiar to AutoCAD users for many years, is graphically incorporated in all sketching exercises as depicted in the following examples.

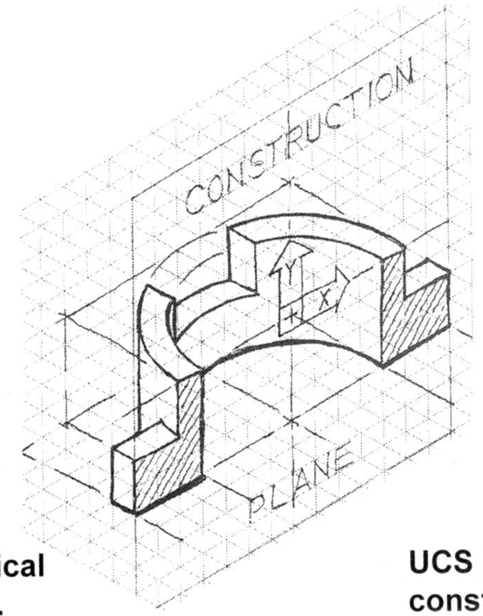

UCS icon on a vertical construction plane.

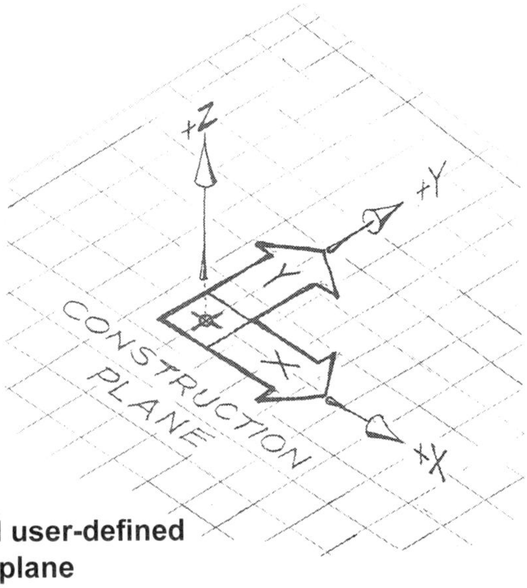

UCS icon and user-defined construction plane

We are sure you will want to use a standard CAD text, and possibly an engineering graphics text, along with this sketching workbook. You may be perplexed as to why 3D modeling is presented first, before 2D conventions. This decision is based on a paradigm acknowledging that engineering design, whether in a designer's mind or put in tangible form, is a 3D exercise. You may be forced to represent a 3D concept on a two-dimensional surface—either in a sketch, drawing, or CAD model. But the thinking must be 3D to start with. This is what we try to instill in each and every student. All sketches become representative of this 3D mental model. The sooner your students start three-dimensional thoughts, the more efficient will be their design process. Because we believe sketching is the vehicle for planning CAD, sketching should lead in the most natural sense to CAD modeling.

To Students

The vast majority of you were good sketchers in kindergarten and first grade. You sketched because it was the most efficient way for you to explain your world. After all the years, you are back at the same point. Sketching is nearly too independent. You can do it anywhere, with almost any material. Sketching is immediate. It involves little investment in time or ego. Spend a few hours on a detailed drawing and your natural inclination is to avoid changing it, even in the face of overwhelming evidence that your design simply doesn't work.

Sketching allows you to be efficient and productive in CAD. Sitting down at a CAD workstation with little conception as to how you are going to solve a problem is the surest way to guarantee that you will be inefficient. Because sketching doesn't have a fixed tool box, you might entertain a solution that, were you to do all your thinking with the computer, you might overlook.

How good a sketcher do you have to be to be a successful visualizer and designer? Your sketches must first communicate to yourself. Many landmark designs began life as seemingly unintelligible squiggles. But beyond communicating with yourself, at some time you must communicate your ideas to coworkers and your boss. At that time, a sketch must communicate its technical and geometric data exactly. This is the level of sketching to which you should aspire.

The examples on the following pages are done in sketch form. The text on the figures and problems is mostly hand-lettered. This was done intentionally to guide your development. If you conscientiously use these as good examples, you should reawaken the sketcher in yourself. You'll be ready to use sketching to visualize and plan your engineering designs, preparing them for eventual CAD completion.

1 Visualization

Coordinated with VMGED Chapter 3

Developing Spatial Skills

A common trait of engineers and technologists is the ability to quickly visualize a problem and its potential solutions. One tool in visualization is freehand sketching—inexpensive, immediate, flexible, and effective. This sketching workbook will guide you through a number of exercises coordinated with *Visualization, Modeling, and Graphics for Engineering Design*. The exercises are intended to improve your ability to visualize spatial relationships so important in engineering design.

Spatial visualization is the process of figuring out where you are in space, how to move in relation to various coordinate axis systems, and how the universe of geometric shapes and engineering practices are represented. As a good visualizer you not only "see" well, you also tend to have a running internal dialogue with yourself. You visualize by saying things such as: "I'm in front of the object," or, "I'm to the right and behind the object." Or possibly, "The hole starts at the top plane and goes half-way to the bottom," or, "The part has been rotated away from me, toward my right shoulder." So go ahead, talk to yourself.

Figure 1-1 shows a number of technical sketching techniques—orthographic and axonometric views; sections and details; notes and dimensions. Each helps you visualize geometry and function.

How do you become an effective visualizer and produce sketches like those in Fig. 1-1? Practice! Practice! Practice! How do you become a good sketcher? Sketch! Sketch! Sketch!

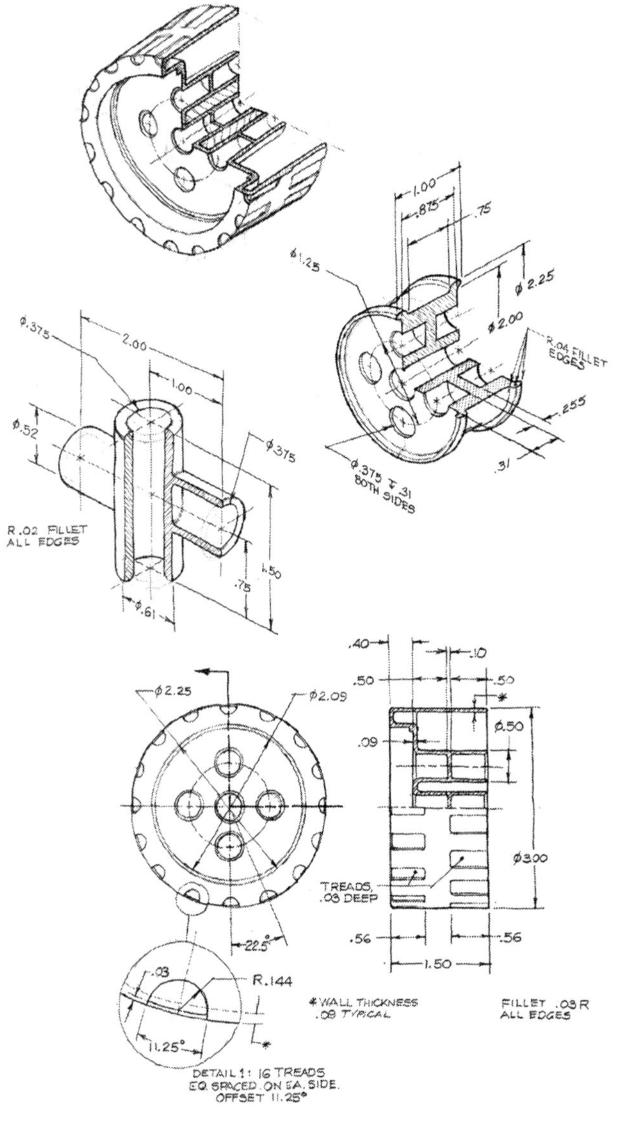

Fig. 1-1. Technical sketches help you visualize and communicate.

Spatial Position

Even though all freehand sketches are two-dimensional because they are executed on a sheet of paper, good visualizers are always thinking in 3D space. The sketching sheet is a construction plane on which a view of three-dimensional space is represented. Two parameters control this representation: viewpoint position, and viewpoint direction.

In orthographic and axonometric views (the views of technical and engineering graphics), **viewpoint position** is actually at infinity, resulting in the phenomenon that parallel features (edges, intersections, and limits) remain parallel. There is no perspective in engineering technical sketching. This parallel condition allows parallel measurements and distances to be compared, an important aspect of technical visualization.

Viewpoint direction can be represented by a vector originating between the viewer's eyes (Fig. 1-2). The vectors are parallel because the viewer is at infinity. The construction plane on which the sketch will be done is perpendicular to these viewing vectors. In this manner, you can plan the direction of the sketch as well as communicate that direction to others.

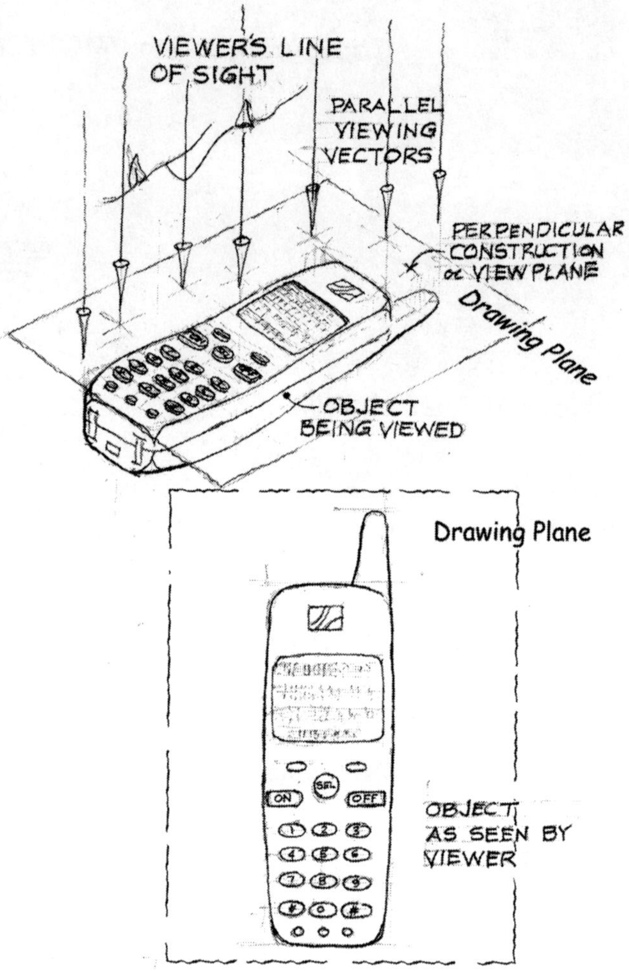

Fig. 1-2. The viewpoint determines the sketching plane and viewing direction.

World and Local Axis Systems

Space is defined by three mutually perpendicular axes centered in the middle of the workspace. Height is measured along the Y-axis; width is measured along the X-axis; depth is measured along the Z-axis. Any point in space can be located by its world coordinates—$X_w Y_w Z_w$. The Z-axis is generally considered to be "in and out" of the paper, the construction plane. Construction (sketching) planes that are perpendicular to the world axes result in what are called "principle" orthographic views—top, front, rear, bottom, and right and left sides (Fig. 1.3). The world axis system is stable. It is not moved or rotated.

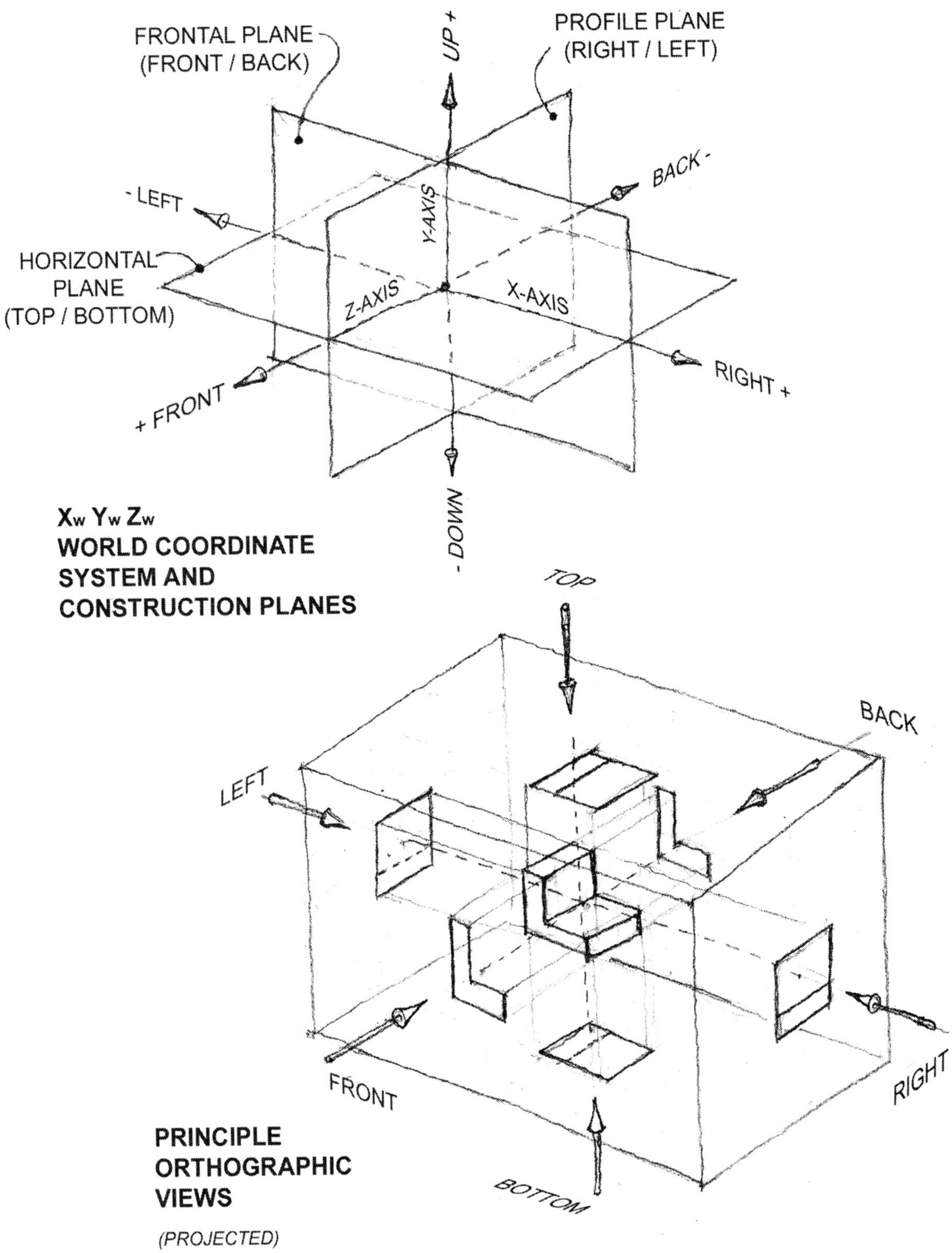

Fig. 1-3. Principle construction planes perpendicular to the world axes.

It is often convenient to orient your sketch to world space, and to orient a central feature of your sketch to the origin of the world axes, $0_w0_w0_w$. You can see a part in Figure 1-4 with the corner of its base located at the world origin. Features on the part, such as the nearby hole, can be easily located relative to the origin. The hole's center is positive Y, negative Z, and is on the X=0 plane.

When an object is removed from the world origin, or it has been rotated so that its features are not aligned with the world axes, it is convenient to establish a local or user-defined axis system ($X_LY_LZ_L$). Figure 1-5 shows the direction of a sketch along a user-defined axis system's Z_L axis. This local axis system is centered on a convenient feature so that other features are easily located. See Fig. 3-7 for an additional example of axis and construction plane relationships. See Chapter 7—Auxiliary Views, for further discussion of auxiliary construction planes.

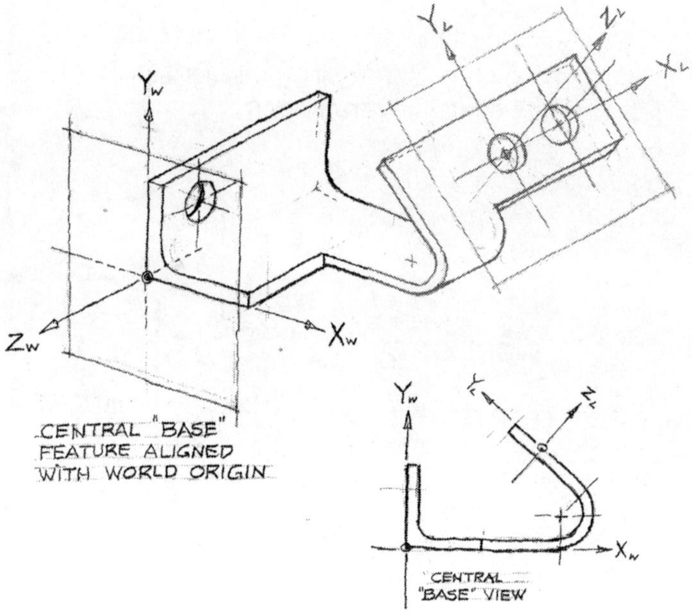

Fig. 1-4. Orient a central feature to the world origin.

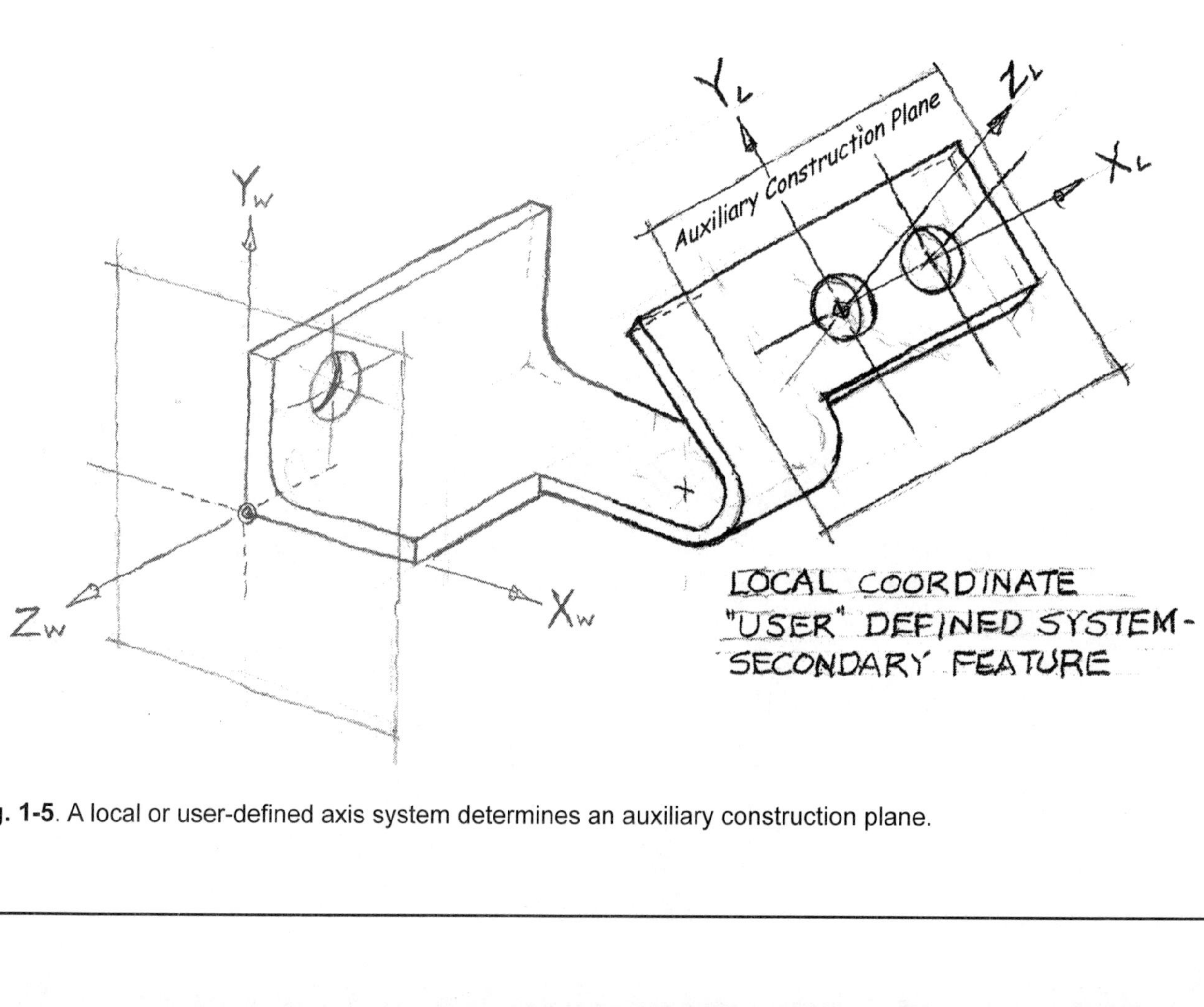

Fig. 1-5. A local or user-defined axis system determines an auxiliary construction plane.

Translation Along Axes

Translation is the operation that moves geometry in space relative (parallel) to an axis system, and is generally shown with a directional arrow. You can translate geometry along one, two, or three axes. You can perform the translations in any order and end up with the same result. Figure 1-6 shows parts that are to be translated into position relative to the Z and Y axes.

The same considerations hold true when translating relative (parallel) to user-defined axes. However, care must be taken when measuring along these non-principle axes. Only when a construction plane is parallel to an axis can measurement be made directly. Translation relative to user-defined axes is shown in Fig. 1-7.

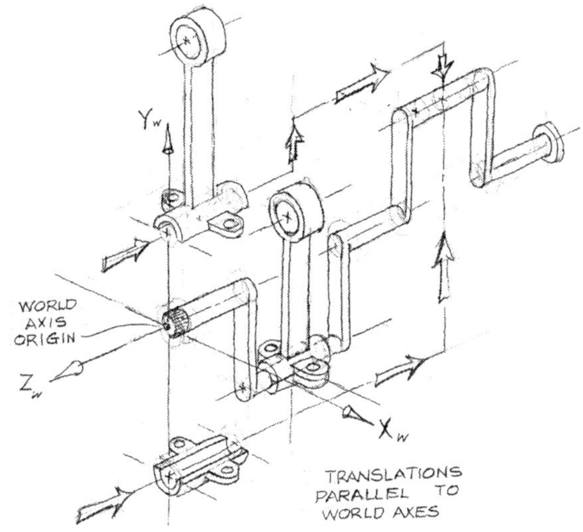

Fig. 1-6. Translation relative to world axes.

Rotation About Axes

Rotation is accomplished in a plane perpendicular to an axis and is represented with a circular arrow. The direction of the arrow shows the direction of rotation (called clockwise or counter-clockwise). Positive rotation adheres to the "right-hand rule." (Place your right hand so your thumb points in the positive direction of the axis. Your fingers will curl in the direction of positive rotation.) Study Fig. 1-8, where positive rotation by the "right-hand rule" is shown for world and user-defined axes.

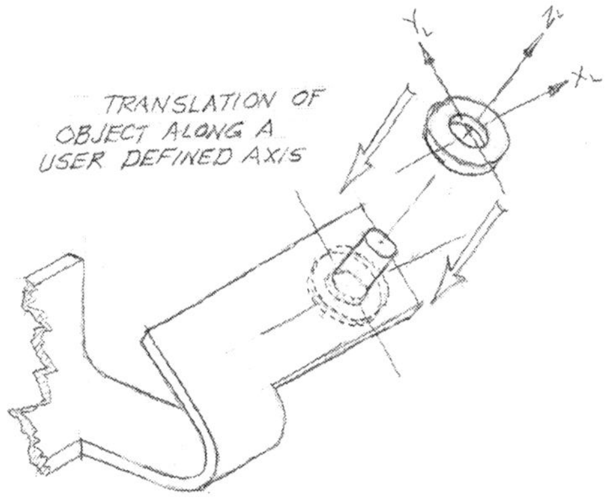

Fig. 1-7. Translation relative to user-defined axes.

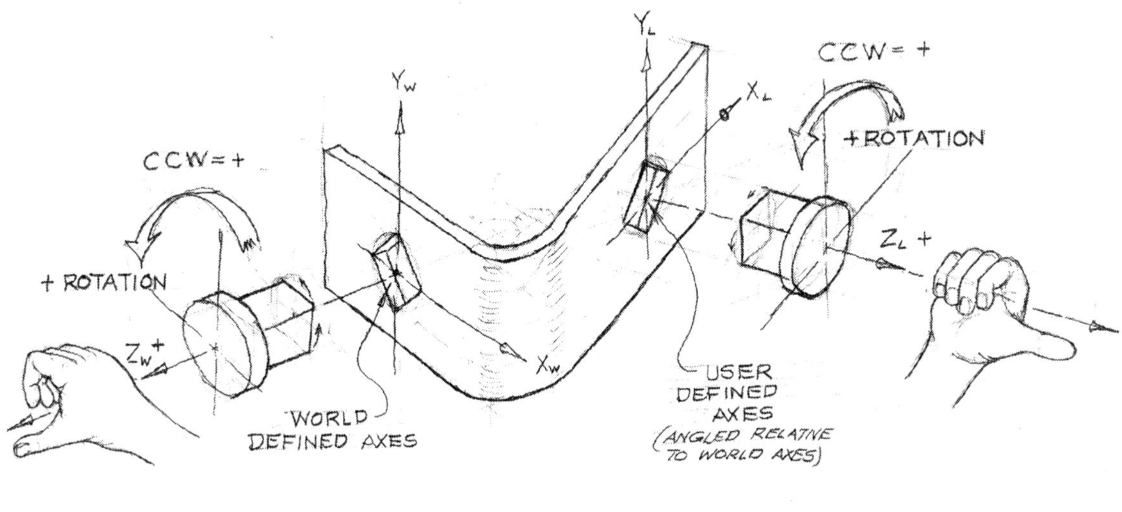

Fig. 1-8. Right hand rotation relative to world and user-defined axes.

Mental Rotation Tasks

Good visualizers are able to mentally rotate objects and determine the result. This ability is helpful in sketching where anticipating what your sketch will look like dramatically increases sketching efficiency. Look at Fig. 1-9. Object 1 is shown in its original and rotated position. Can you determine the correct final position of Object 2 from its original position (C) if it were rotated identically? If you can, do you know how you did it?

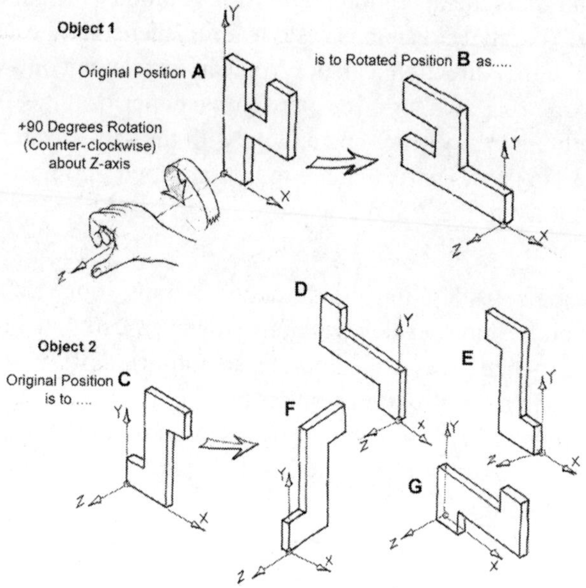

Fig. 1-9. Mental rotation task.

Cutting Plane Tasks

Good visualizers are also able to anticipate the result of a plane passing through an object. This is important because cutting planes are a fundamental construction and visualization technique. If in Fig. 1-10 a plane passes through the object as shown, can you quickly choose the correct result? If you can, do you know how you did it?

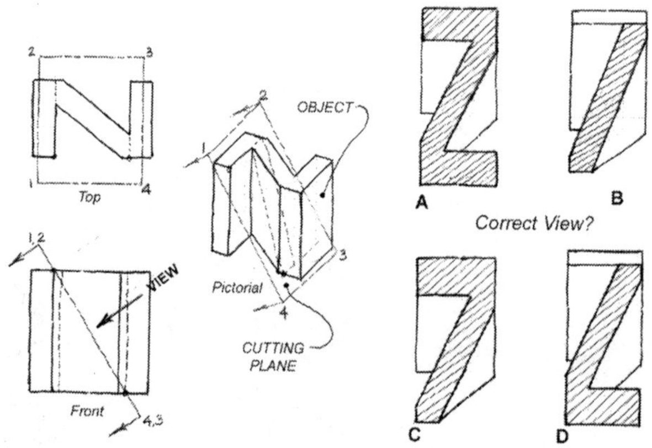

Fig. 1-10. Cutting plane task.

Develop Your Sketching Skills

Many times designers don't really know what they are envisioning until they see it in tangible form. A sketch reinforces the idea and promotes change, refinement, and testing. It is an iterative process—you do it over and over, refining your solution with each subsequent sketch. For that reason, preliminary sketches are usually not works of art. In fact, technical sketches may *never* be works of art, having a level of sophistication only sufficient to communicate the necessary information.

The visualization exercises that follow do not require you to directly sketch objects. Instead, they are designed to give you valuable experience with mental rotation and cutting plane tasks similar to those described in Figs. 1-9 and 1-10.

As you progress through the chapters you'll find it easier to quickly make sketches that effectively communicate your intent. You may even start to "think" with your pencil, using it as an extension of your mind, and if you do, you'll be right where we want you to be.

MENTAL ROTATION: With respect to the X-Y-Z coordinate systems shown in the following problems, the original position of view A is related to the rotated position of B as the original position of view C is related to the rotated position of view D, E, F, or G? Circle the letter D, E, F, or G for the correct answer.

Problem 1.1a

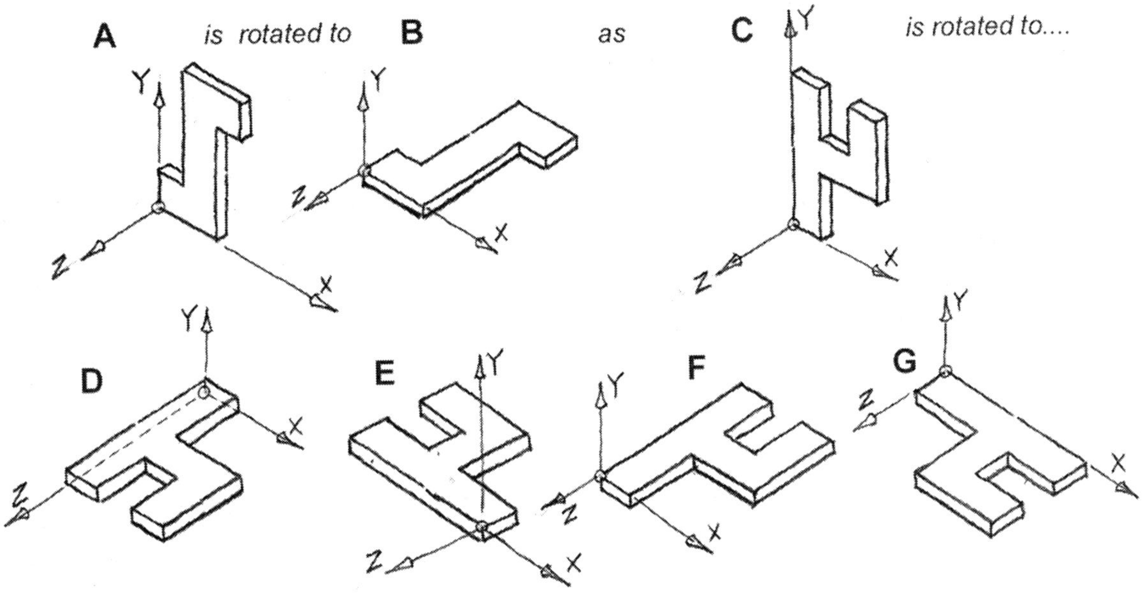

Problem 1.1b

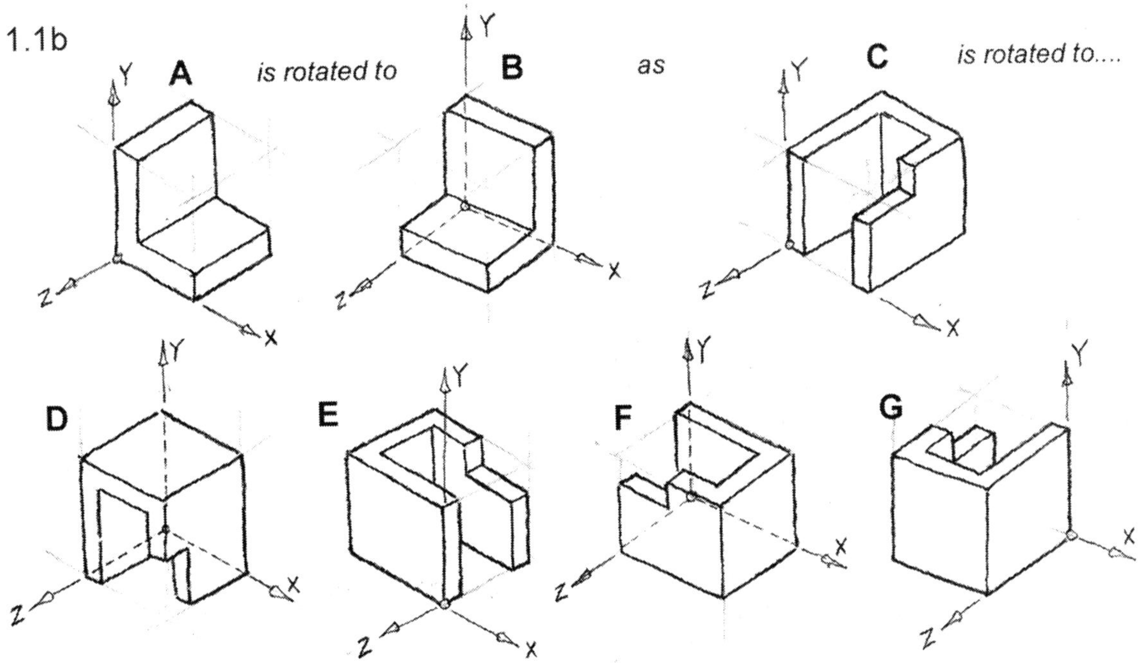

Drawn By:		FREEHAND SKETCHING	
	Scale	TITLE MENTAL ROTATION	
Date	Sheet	Drawing Number A	1.1

MENTAL ROTATION: With respect to the X-Y-Z coordinate systems shown in the following problems, the original position of view A is related to the rotated position of B as the original position of view C is related to the rotated position of view D, E, F, or G? Circle the letter D, E, F, or G for the correct answer.

Problem 1.2a

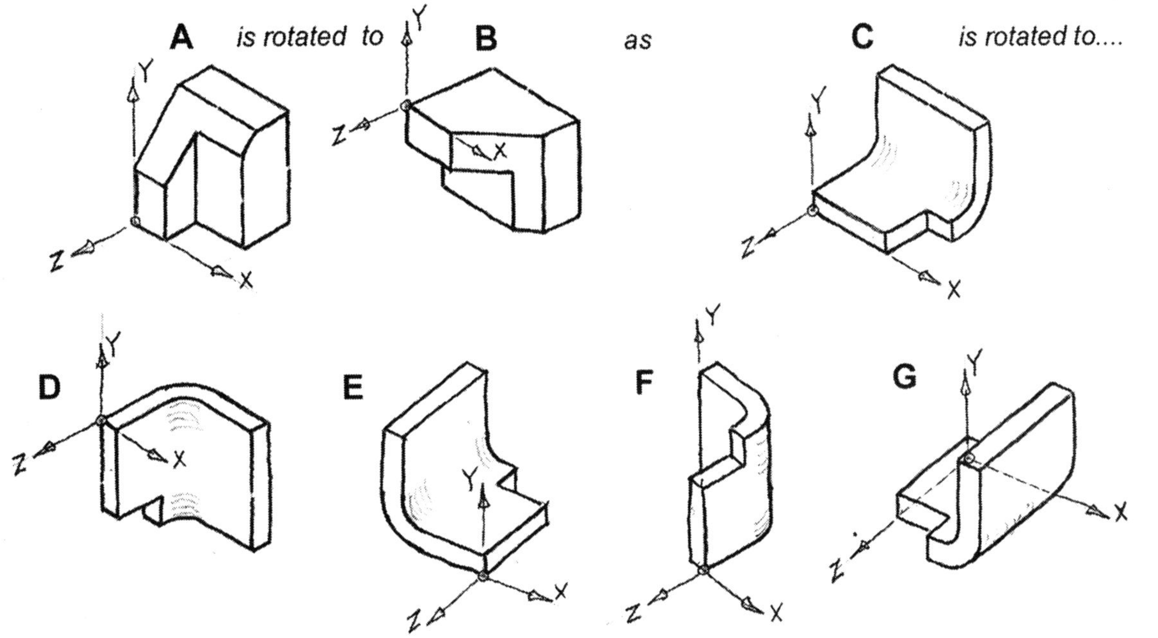

Problem 1.2b

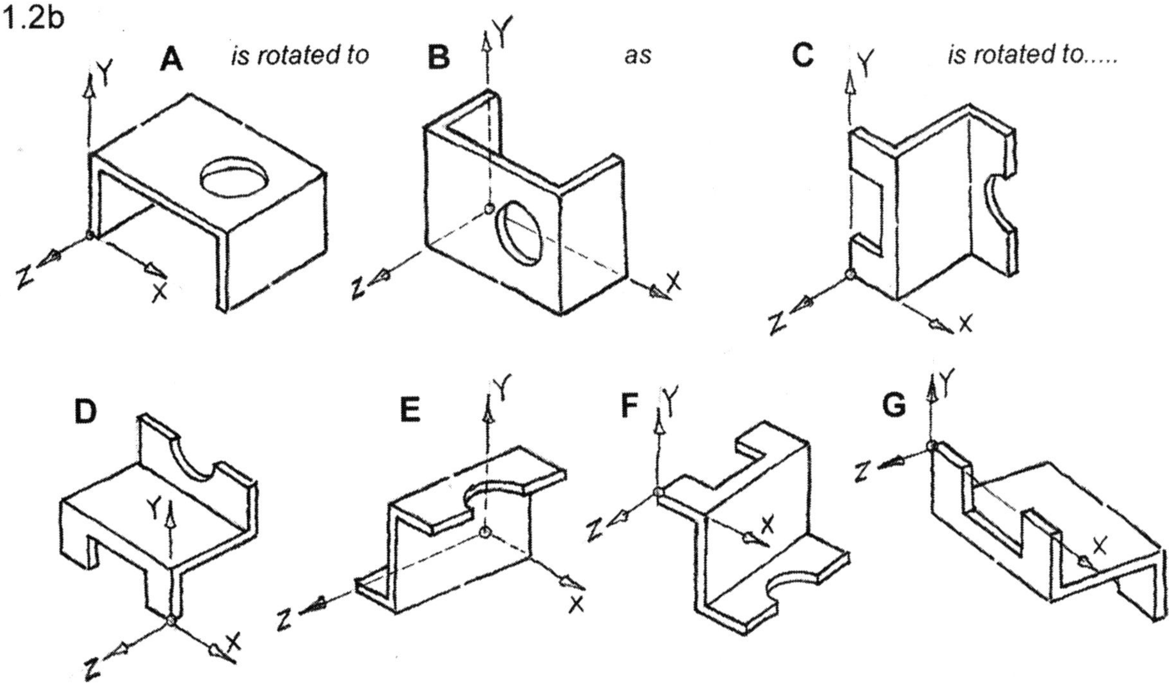

Drawn By:		FREEHAND SKETCHING	
	Scale	TITLE MENTAL ROTATION	
Date	Sheet	Drawing Number A	1.2

MENTAL ROTATION: With respect to the X-Y-Z coordinate systems shown in the following problems, the original position of view A is related to the rotated position of B as the original position of view C is related to the rotated position of view D, E, F, or G? Circle the letter D, E, F, or G for the correct answer.

Problem 1.3a

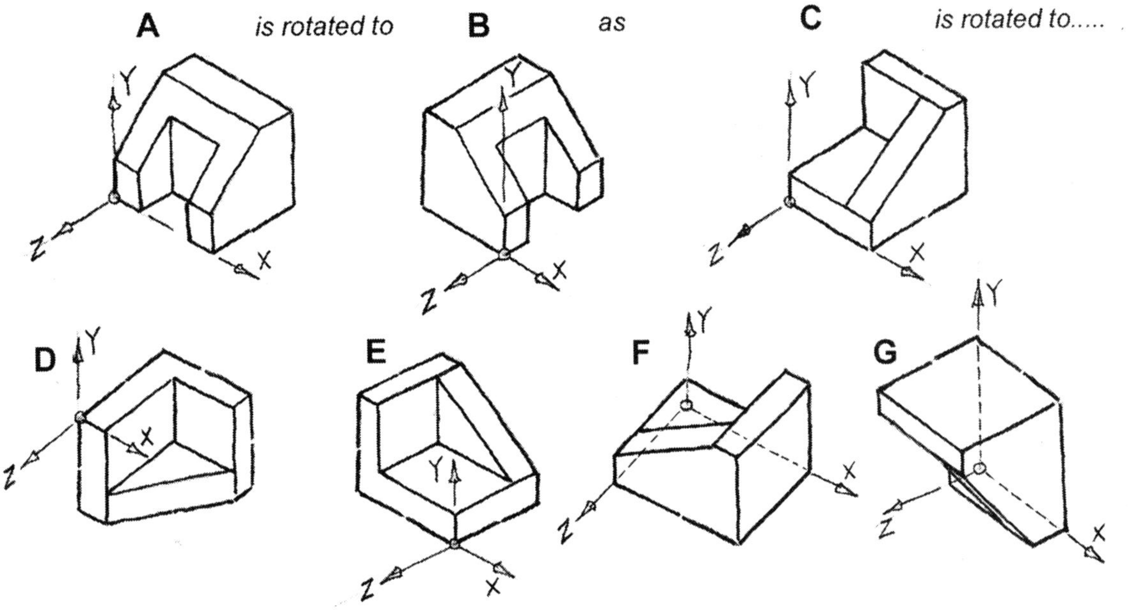

Problem 1.3b

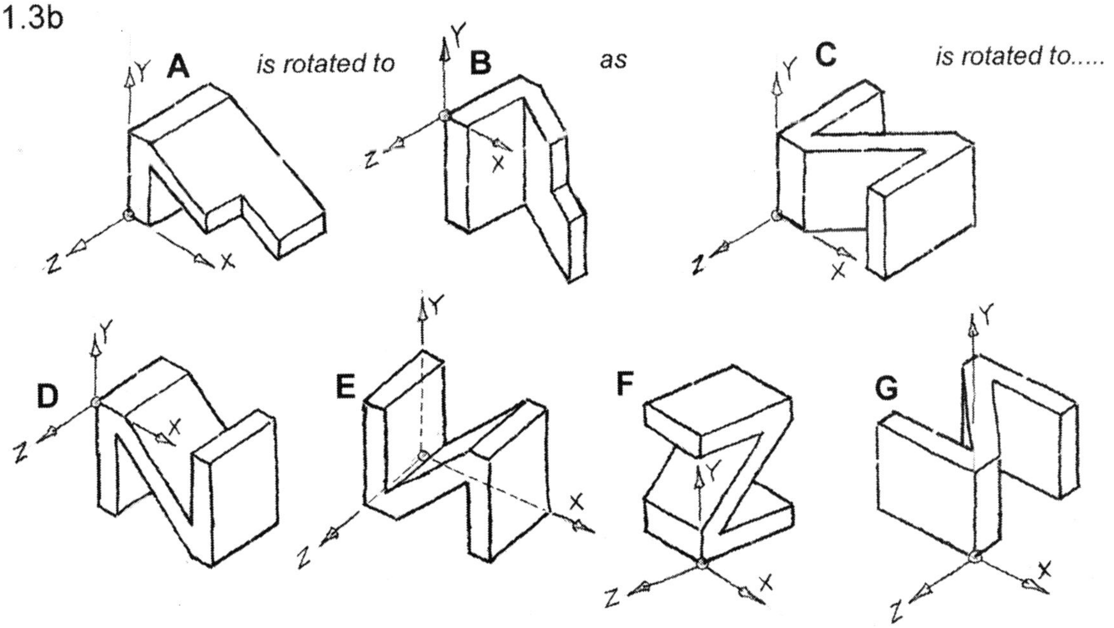

Drawn By:		FREEHAND SKETCHING	
	Scale	TITLE MENTAL ROTATION	
Date	Sheet	Drawing Number A	1.3

MENTAL ROTATION: With respect to the X-Y-Z coordinate systems shown in the following problems, the original position of view A is related to the rotated position of B as the original position of view C is related to the rotated position of view D, E, F, or G? Circle the letter D, E, F, or G for the correct answer.

Problem 1.4a

A *is rotated to* B *as* C *is rotated to.....*

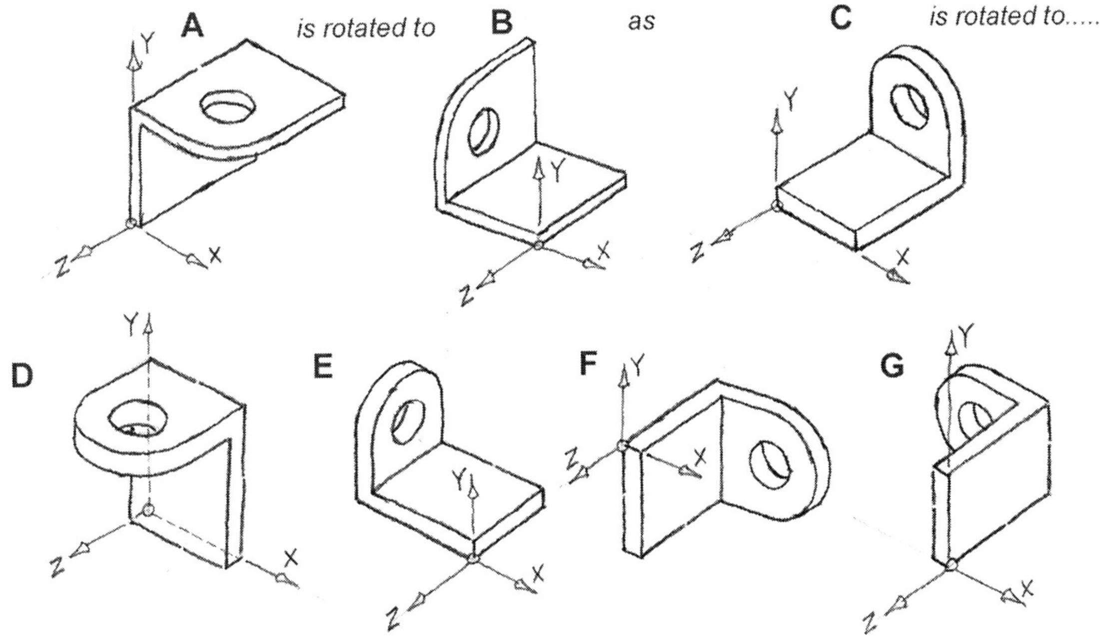

Problem 1.4b

A *is rotated to* B *as* C *is rotated to.....*

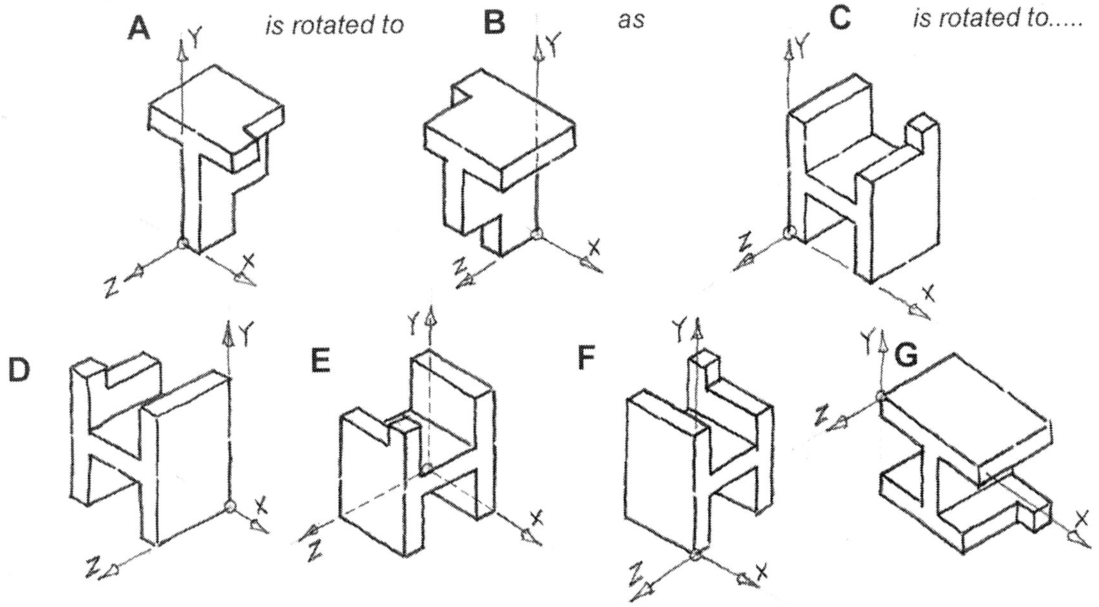

Drawn By:		FREEHAND SKETCHING	
	Scale	TITLE MENTAL ROTATION	
Date	Sheet	Drawing Number A	1.4

MENTAL ROTATION: With respect to the X-Y-Z coordinate systems shown in the following problems, the original position of view A is related to the rotated position of B as the original position of view C is related to the rotated position of view D, E, F, or G? Circle the letter D, E, F, or G for the correct answer.

Problem 1.5a

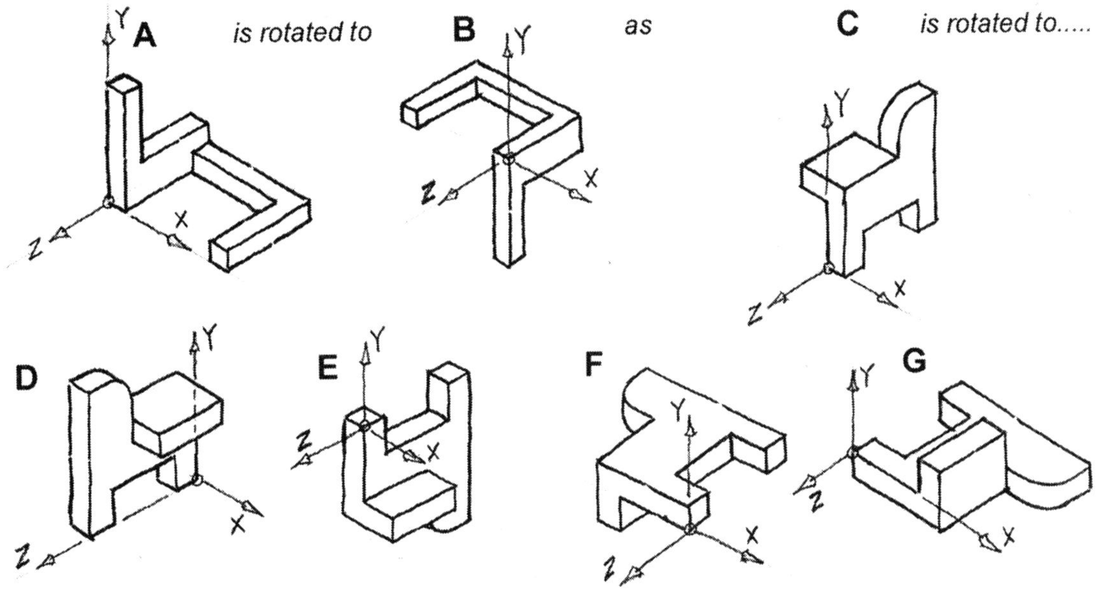

Problem 1.5b

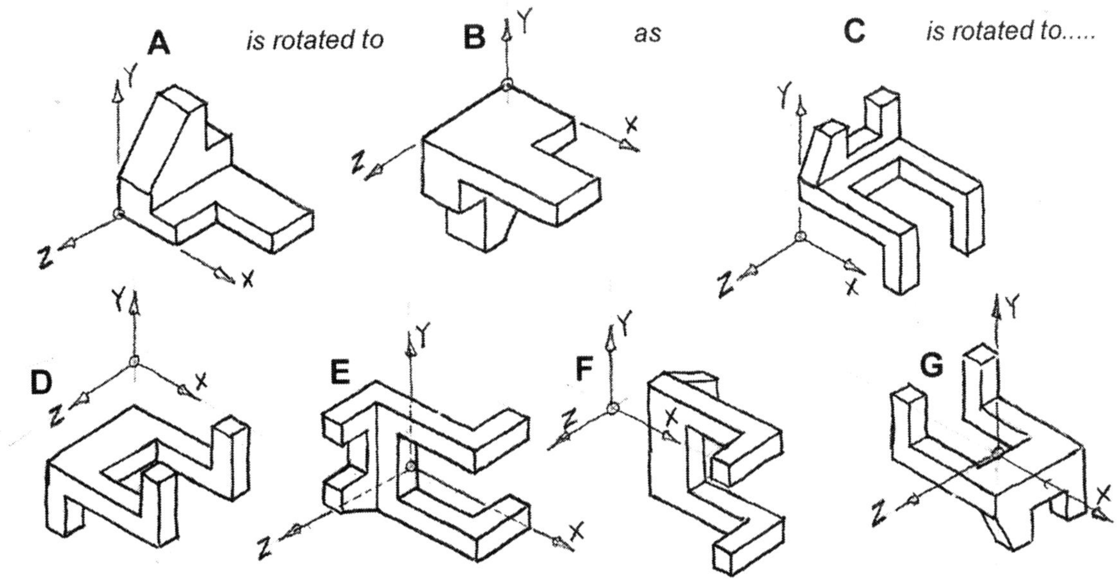

Drawn By:		FREEHAND SKETCHING	
	Scale	TITLE MENTAL ROTATION	
Date	Sheet	Drawing Number A	1.5

CUTTING PLANES: The intersection of a cutting plane and a 3D solid object results in a cut-away or cross-sectional area. The cross-section area is a two-dimensional shape whose edges are defined by the boundaries of the surfaces of the original 3D solid object.

Cutting Plane Problems: Study the objects and cutting planes shown below (identified by corners 1-2-3-4). Refer to page 1.10 and enter the letter of the correct matching cutting plane view.

Problem 1.6a

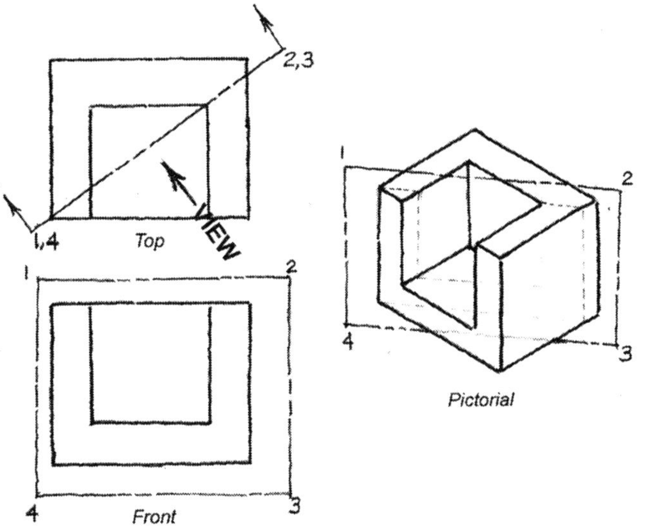

Matching View ____
(from p. 1.10)

Problem 1.6b

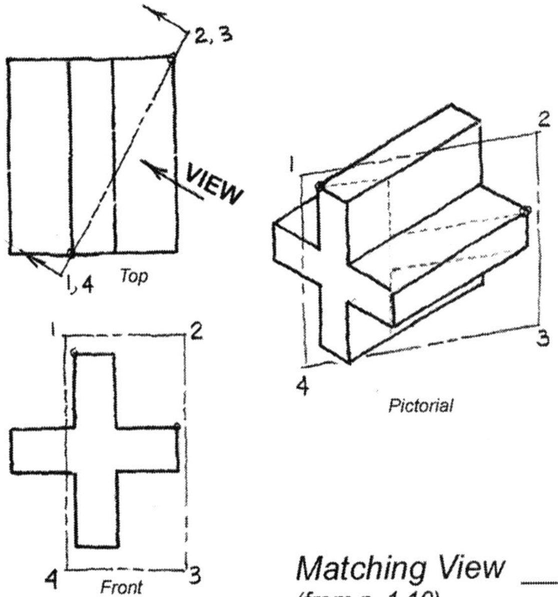

Matching View ____
(from p. 1.10)

Problem 1.6c

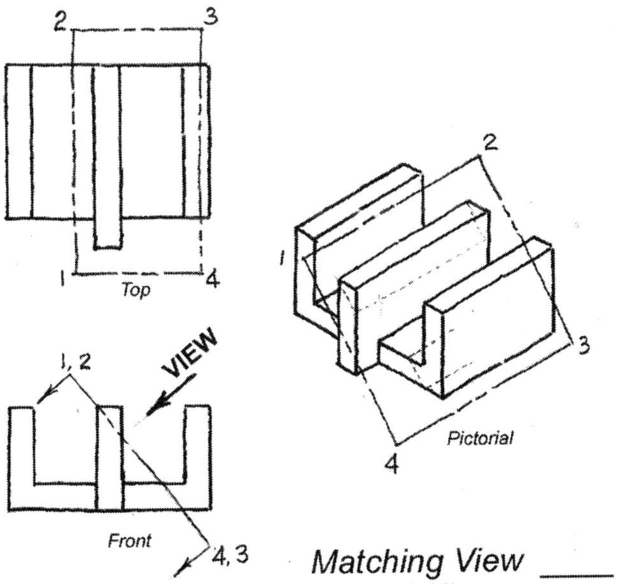

Matching View ____
(from p. 1.10)

Problem 1.6d

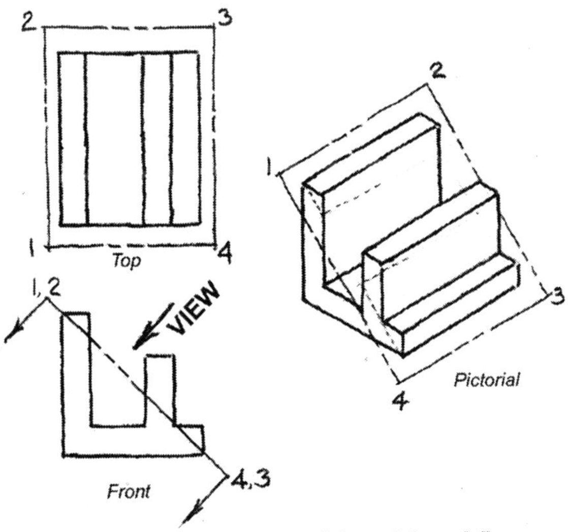

Matching View ____
(from p. 1.10)

	Drawn By:		VISUALIZATION	
		Scale	TITLE CUTTING PLANES	
	Date	Sheet	Drawing Number A	1.6

CUTTING PLANES: The intersection of a cutting plane and a 3D solid object results in a cut-away or cross-sectional area. The cross-section area is a two-dimensional shape whose edges are defined by the boundaries of the surfaces of the original 3D solid object.

Cutting Plane Problems: Study the objects and cutting planes shown below (identified by corners 1-2-3-4). Refer to page 1.10 and enter the letter of the correct matching cutting plane view.

Problem 1.7a

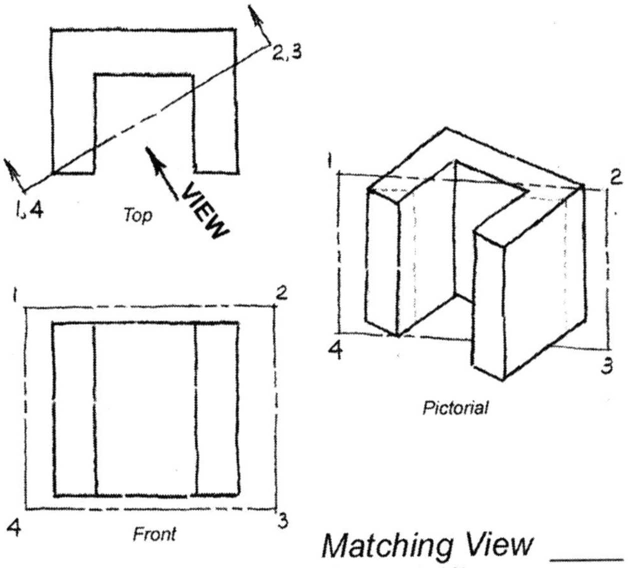

Matching View ____
(from p. 1.10)

Problem 1.7b

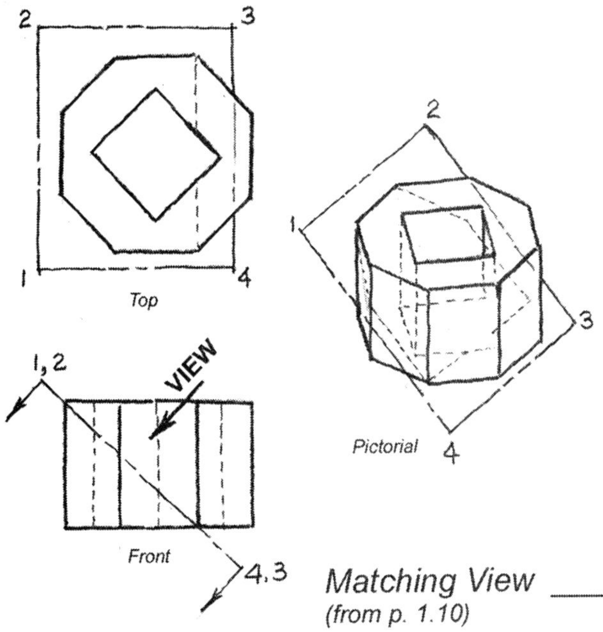

Matching View ____
(from p. 1.10)

Problem 1.7c

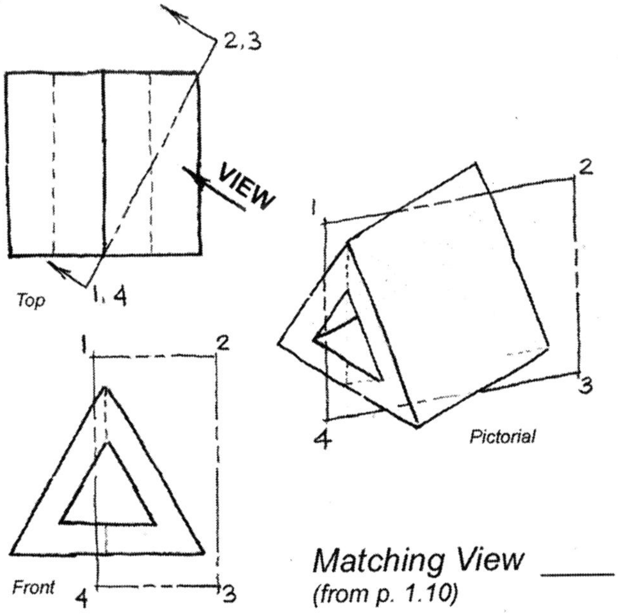

Matching View ____
(from p. 1.10)

Problem 1.7d

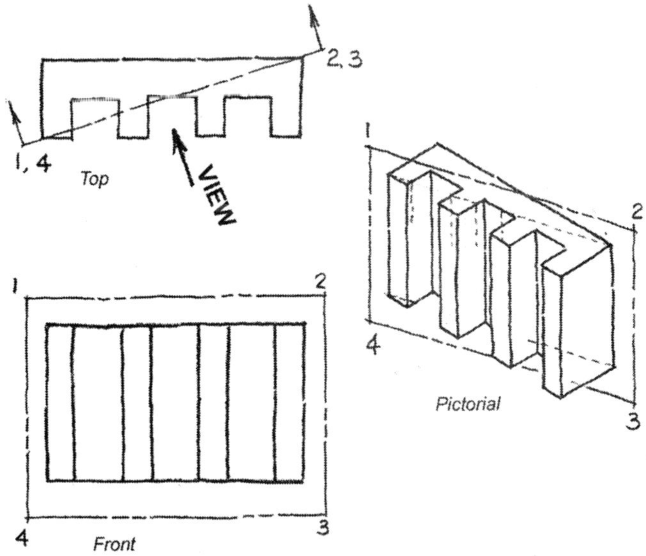

Matching View ____
(from p. 1.10)

Drawn By:		FREEHAND SKETCHING	
	Scale	TITLE CUTTING PLANES	
Date	Sheet	Drawing Number A	1.7

CUTTING PLANES: The intersection of a cutting plane and a 3D solid object results in a cut-away or cross-sectional area. The cross-section area is a two-dimensional shape whose edges are defined by the boundaries of the surfaces of the original 3D solid object.

Cutting Plane Problems: Study the objects and cutting planes shown below (identified by corners 1-2-3-4). Refer to page 1.10 and enter the letter of the correct matching cutting plane view.

Problem 1.8a

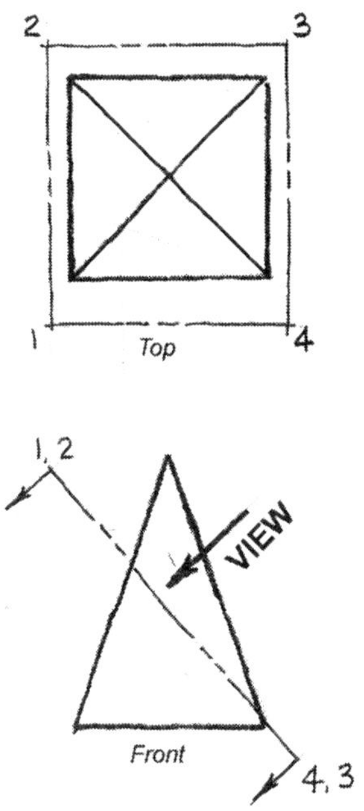

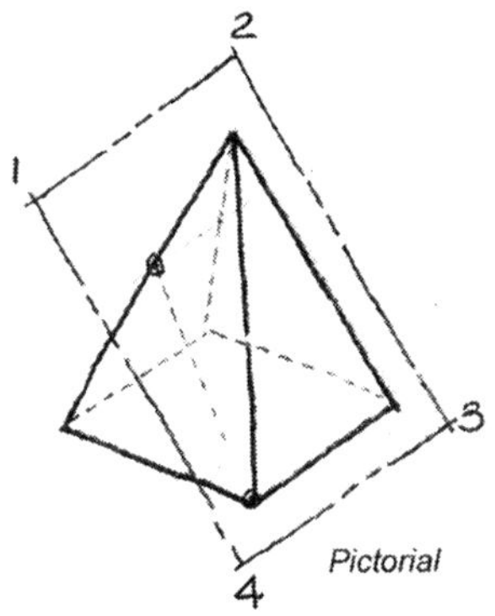

Matching View ____
(from p. 1.10)

Problem 1.8b

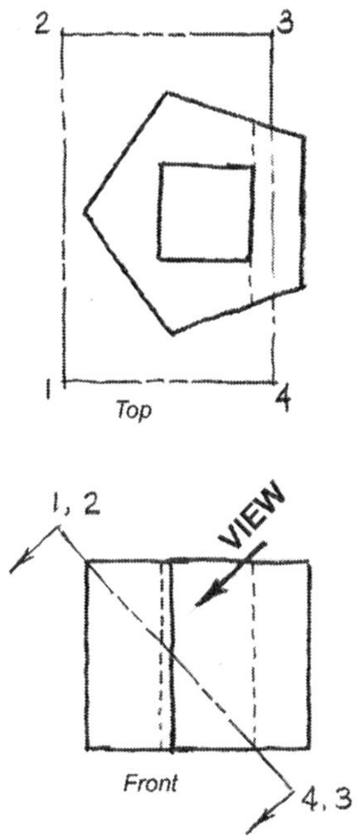

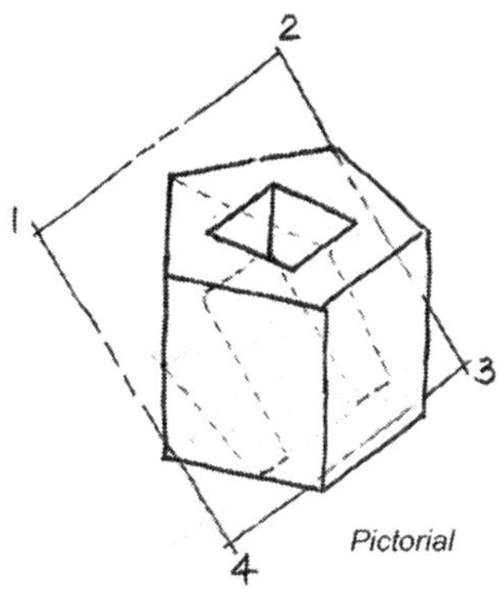

Matching View ____
(from p. 1.10)

Problem 1.8c

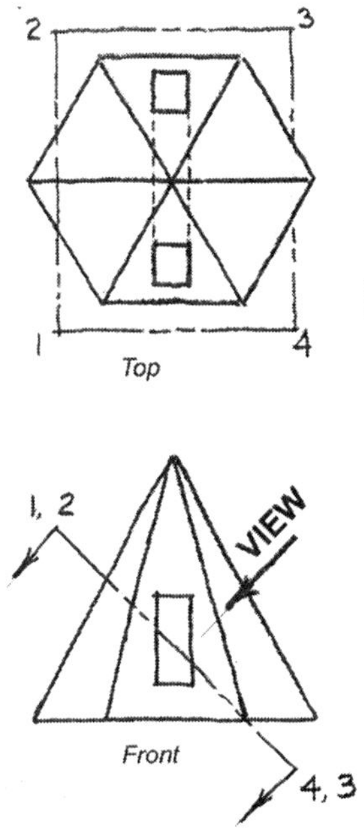

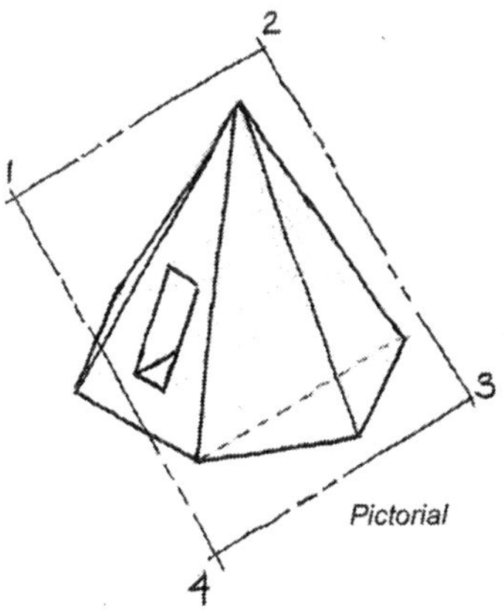

Matching View ____
(from p. 1.10)

Problem 1.8d

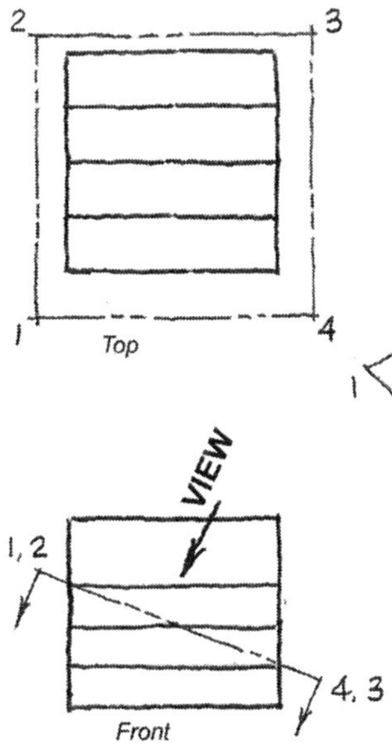

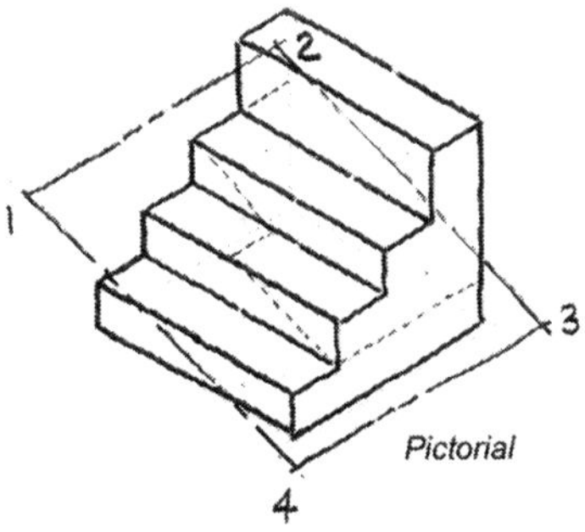

Matching View ____
(from p. 1.10)

Drawn By:		FREEHAND SKETCHING	
	Scale	TITLE CUTTING PLANES	
Date	Sheet	Drawing Number A	1.8

CUTTING PLANES: The intersection of a cutting plane and a 3D solid object results in a cut-away or cross-sectional area. The cross-section area is a two-dimensional shape whose edges are defined by the boundaries of the surfaces of the original 3D solid object.

Cutting Plane Problems: Study the objects and cutting planes shown below (identified by corners 1-2-3-4). Refer to page 1.10 and enter the letter of the correct matching cutting plane view.

Problem 1.9a

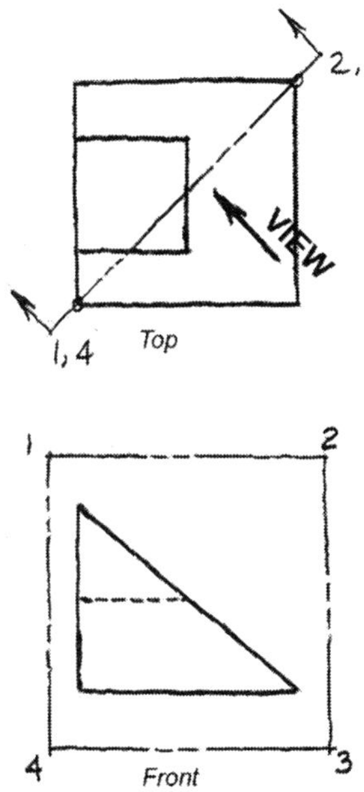

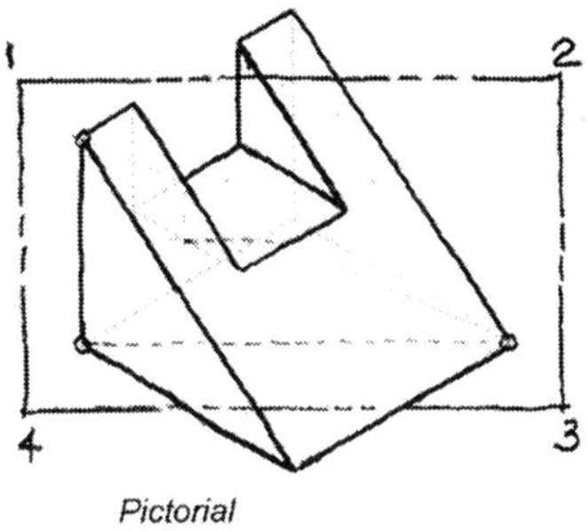

Matching View ____
(from p. 1.10)

Problem 1.9b

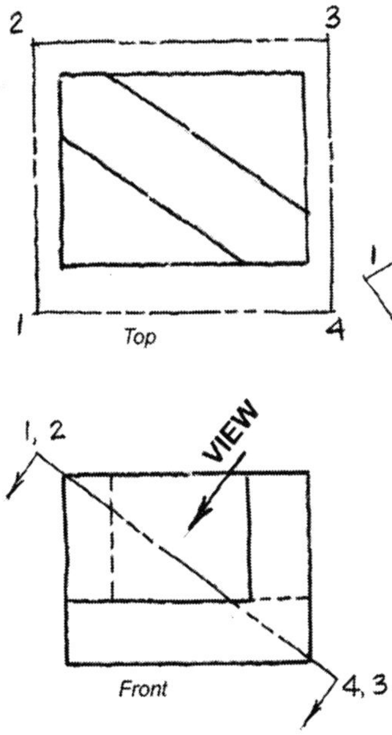

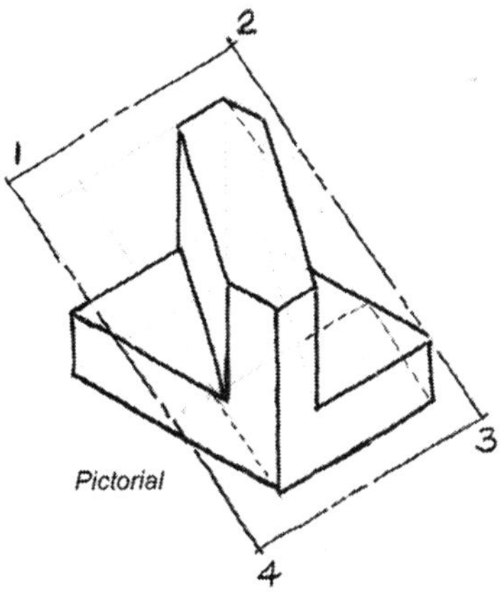

Matching View ____
(from p. 1.10)

Problem 1.9c

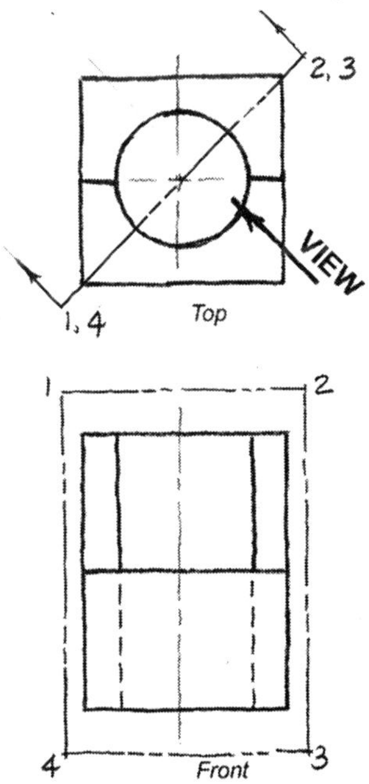

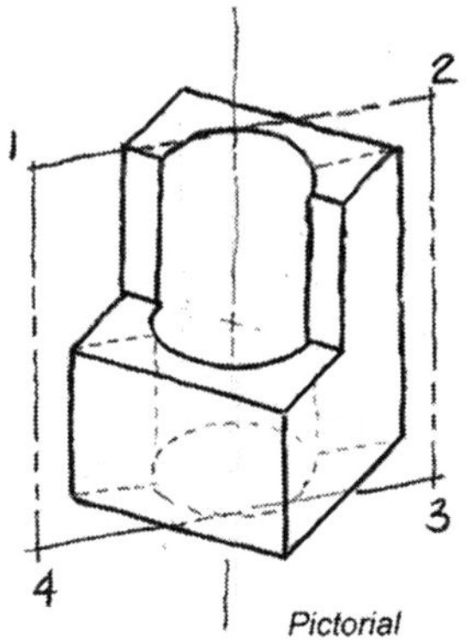

Matching View ____
(from p. 1.10)

Problem 1.9d

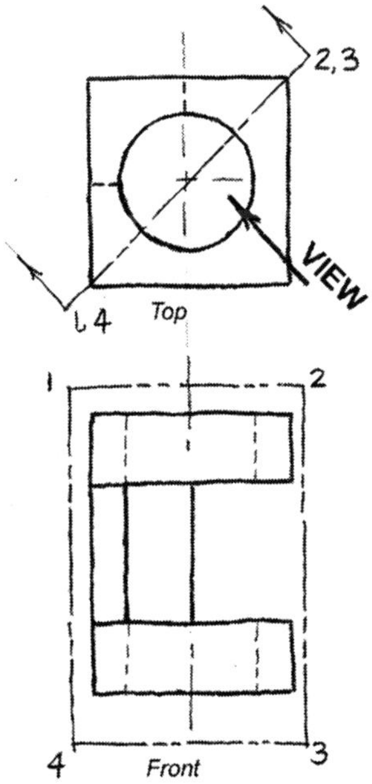

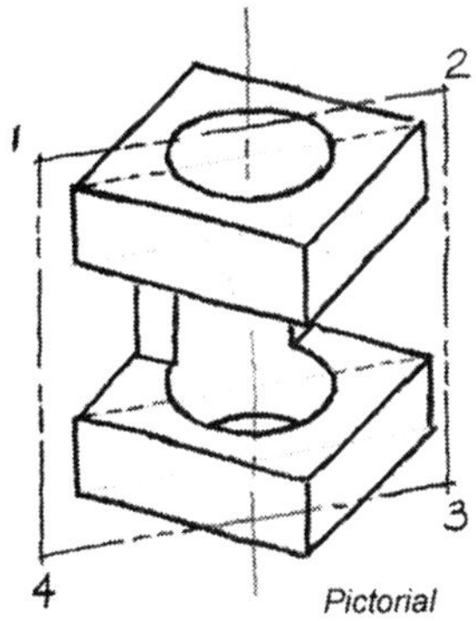

Matching View ____
(from p. 1.10)

	Drawn By:		FREEHAND SKETCHING	
		Scale	TITLE CUTTING PLANES	
	Date	Sheet	Drawing Number A	1.9

CUTTING PLANES: The intersection of a cutting plane and a 3D solid object results in a cut-away or cross-sectional area. The cross-section area is a two-dimensional shape whose edges are defined by the boundaries of the surfaces of the original 3D solid object.

Table of Cutting Plane Views: Based on the problems shown on pages 1.6 - 1.9, find the correct matching view from the view views shown below and label the view on that problem.

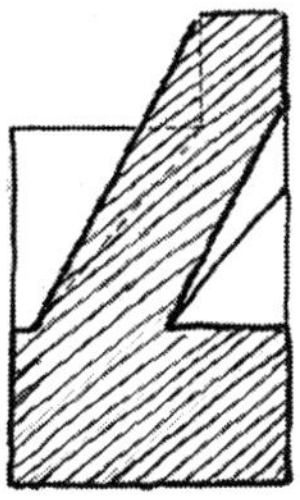

A

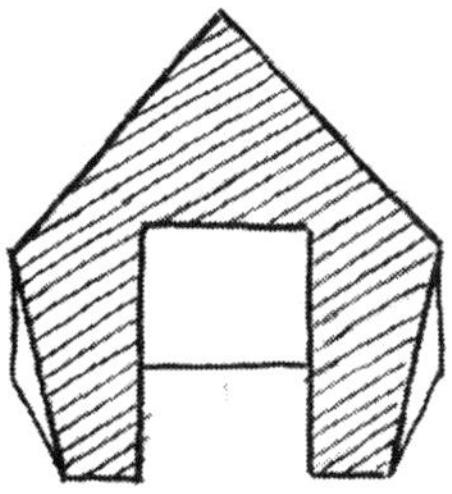

B

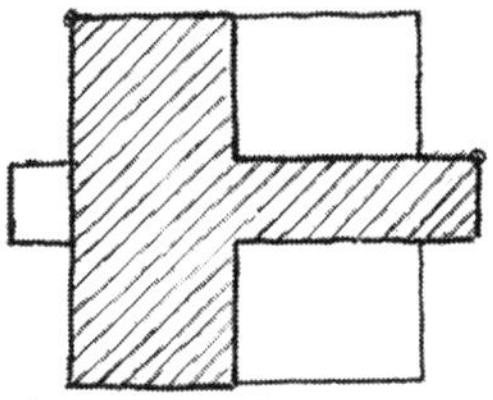

C

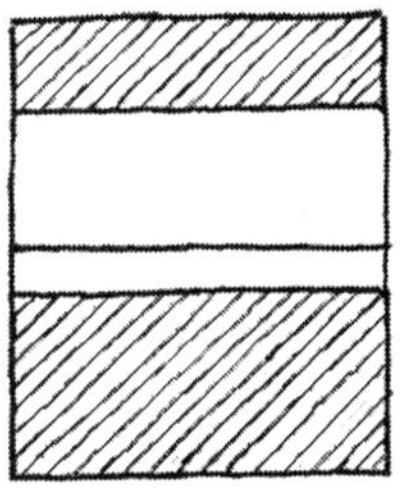

D

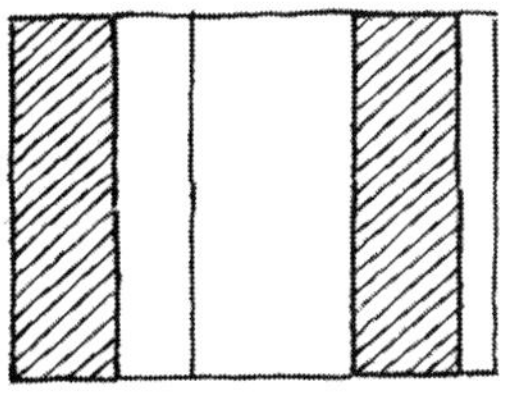

E

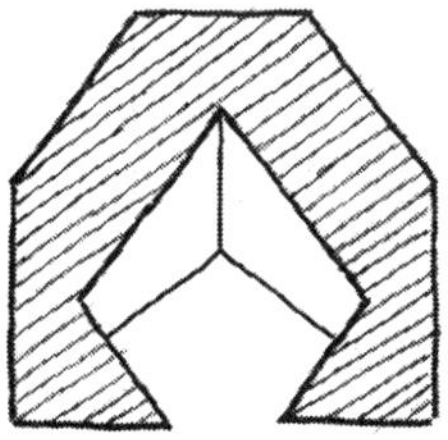

F

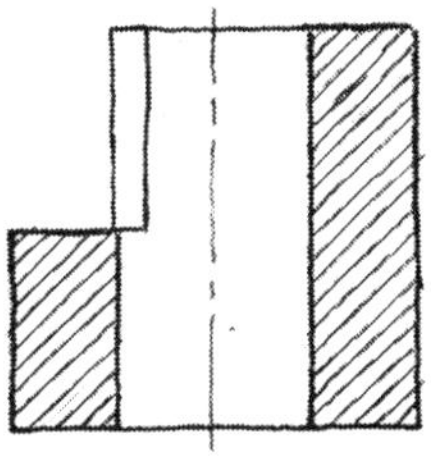

G

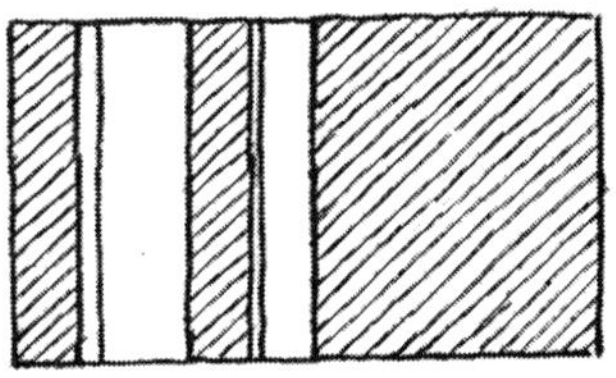

H

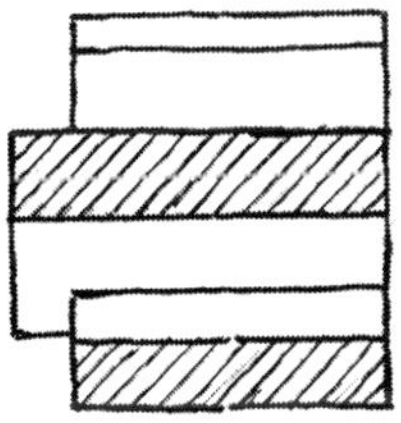

J

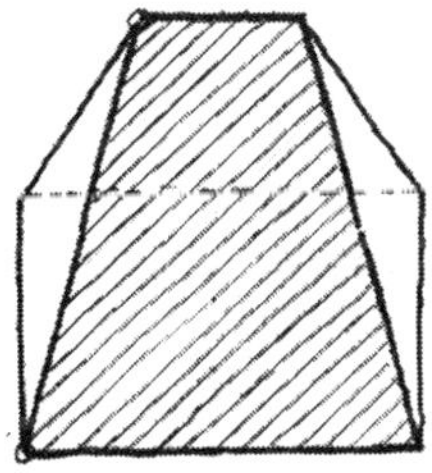

K

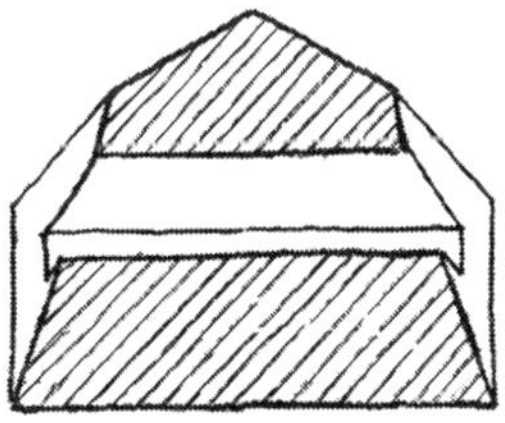

L

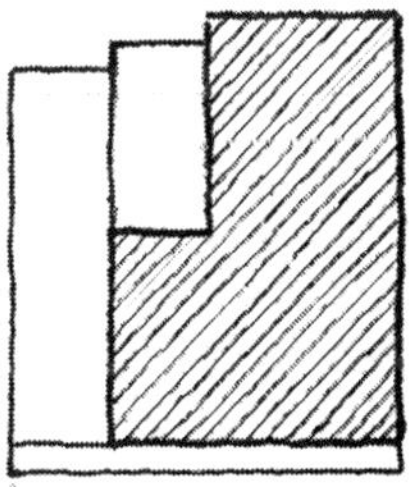

M

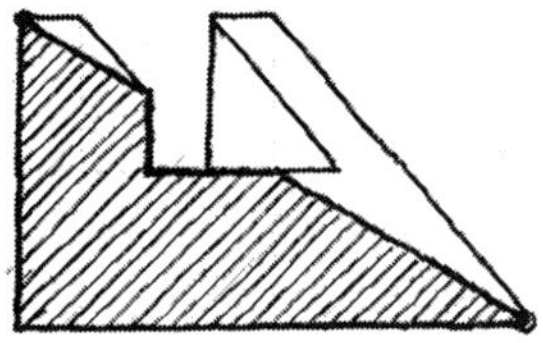

N

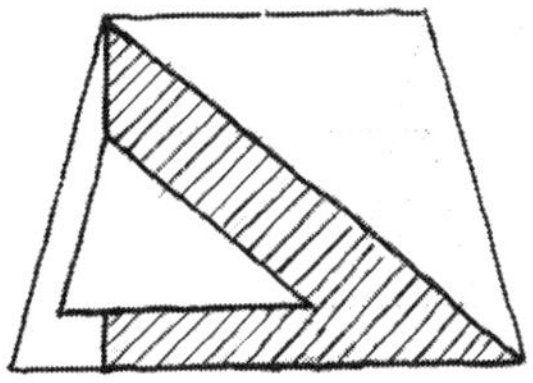

P

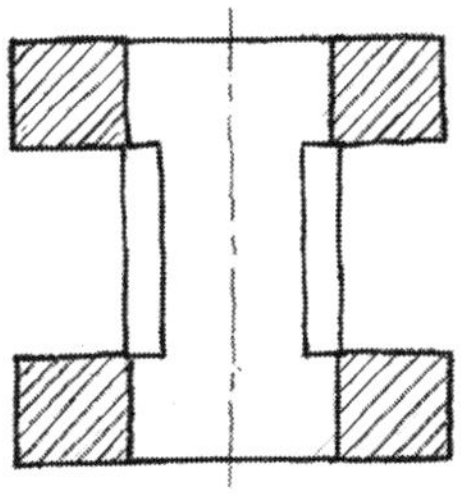

Q

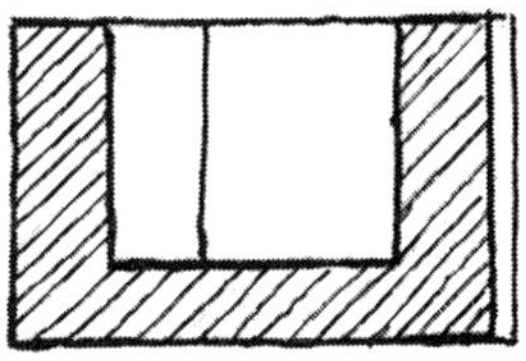

R

Drawn By:		FREEHAND SKETCHING	
	Scale	TITLE CUTTING PLANE VIEWS	
Date	Sheet	Drawing Number A	1.10

2 Spatial Sketching

Coordinated with VMGED Chapter 2

Elements of a Technical Sketch

Sketching was presented in Chapter 1 as a means of understanding and exploring spatial relationships. Certain spatial abilities such as rotations, translations, and mental cutting were presented as characteristic of an effective visualizer. We also pointed out that sketching was an important tool in visualization. But what makes a "good" sketch?

A good sketch communicates the designer's intent in an unambiguous manner. To do that, certain conventions need to be practiced. This includes how you draw and represent space. Three elements comprise a technical sketch: representation of spatial relationships, accompanying text, and graphical designations of movement, rotation, and order.

Cues to Visualizing Space

There are cues that help you determine spatial relationships. Study Fig. 2-1 where several cues to space have been employed. Without these cues, the sketch would not effectively communicate form or position. In nature you have several cues that are not generally employed in technical sketches because of the orthographic nature of the projections and the types of graphical techniques employed for showing materials. In orthographic projections, a distant object is not any smaller than the same object closer to the viewer. They are the same size. And, the color and detail of distant objects are not influenced by the atmosphere (think of mountains as they recede into the distance).

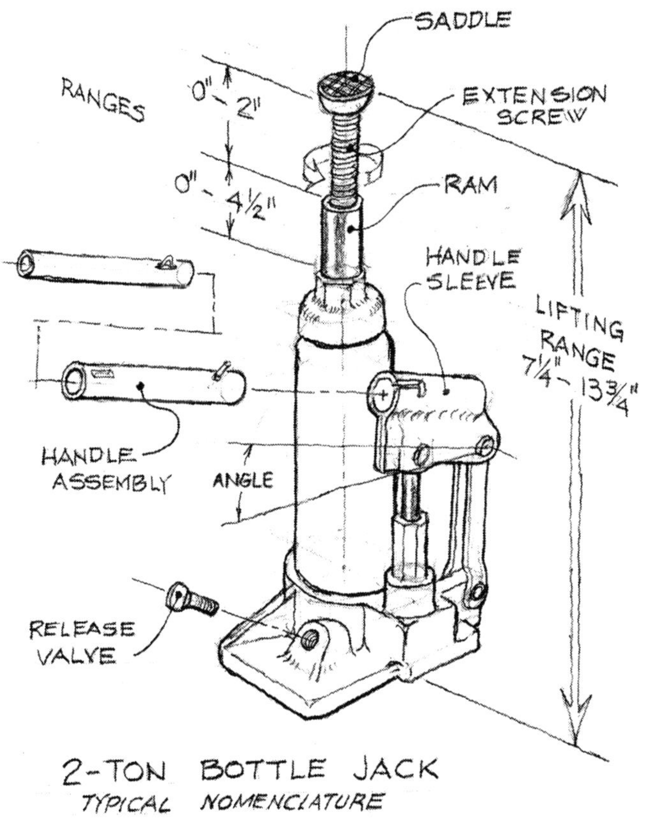

Fig. 2-1. A sketch uses cues to visualizing space.

The first cue to space in a sketch is to use grid paper where real-world dimensions can be related to a grid interval. It may be helpful to show a scale on the sketch to give a sense of size (Fig. 2-2).

Next is to represent mutually perpendicular axis directions of the sketch space with an axis icon (Fig. 2-3). This allows you to determine relationships such as: "how far above, how far to the left, and how far below."

Because objects in orthographic space do not appear smaller as they recede into the distance, you need to employ other cues. This is especially important in principle orthographic views—front, top, rear, etc. Take for example the three objects in Fig. 2-4(a). The cylinder is in front of the prism and the prism is in front of the pyramid. How do you know this? The spatial arrangement, the juxtaposition of the objects, includes overlap. This is your first cue to spatial position. However, in Fig. 2-4(b), this overlap has been removed and it is now impossible to determine which object is in front because the construction plane's Z-axis is "in and out" of the paper.

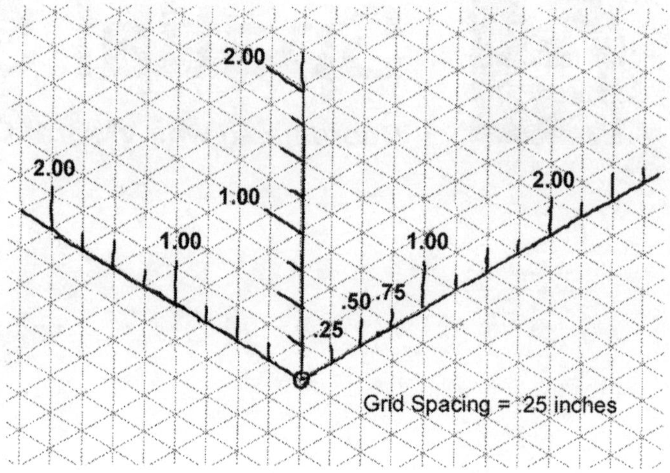

Fig. 2-2. A grid establishes sketch scale.

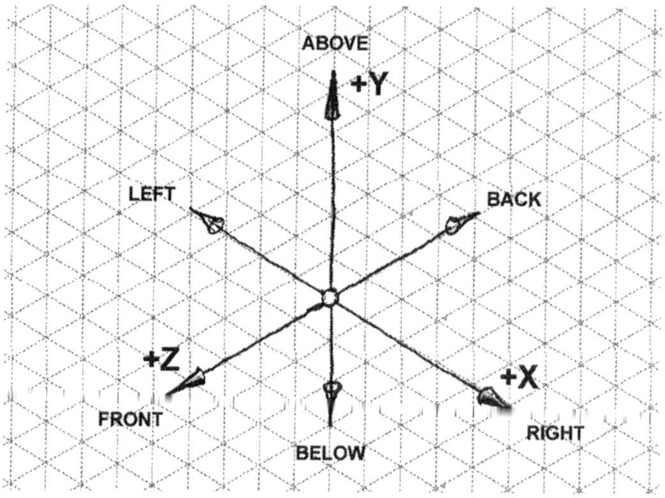

Fig. 2-3. An axis icon establishes spatial direction.

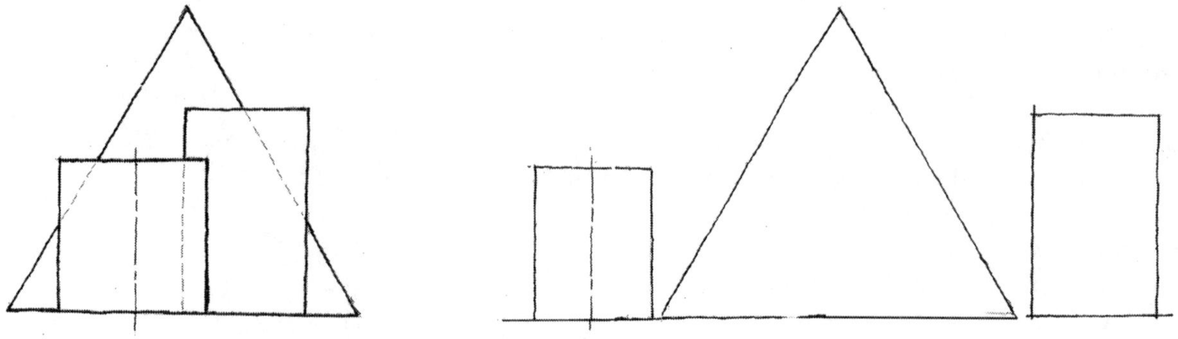

(a) (b)

Fig. 2-4. Overlap helps you visualize spatial position.

Notice in Fig. 2-5 that a view in a direction parallel to the Y-axis has been added, aligned in such a manner that the depth position of the three geometric objects can be determined. As you look between the views you are "reading" the depth and in doing so, you are visualizing their positions in space.

You can also help visualize space and geometry by the type of sketch lines you make. This is where a little "artistic" ability is helpful but it is a technique that can be learned. Simply make the outside lines of an object's profile darker and bolder; make the inside lines showing edges and intersections thinner and lighter; make smooth transitions even thinner and broken. You can see in Fig. 2-6 that the use of varying line weights can aid in visualizing geometry.

This leads to the visualization power of showing all three principle world axes in the form of a pictorial sketch. Instead of viewing space parallel to one of the world axes, look down from an angle. Notice how much easier it is to show spatial position in the pictorial view of Fig. 2-7 than in Figs. 2-4 or 2-5. This is why most assembly instructions include a pictorial representation of the parts. You may also recognize the use of the axis icon, overlap, and varying sketch lines.

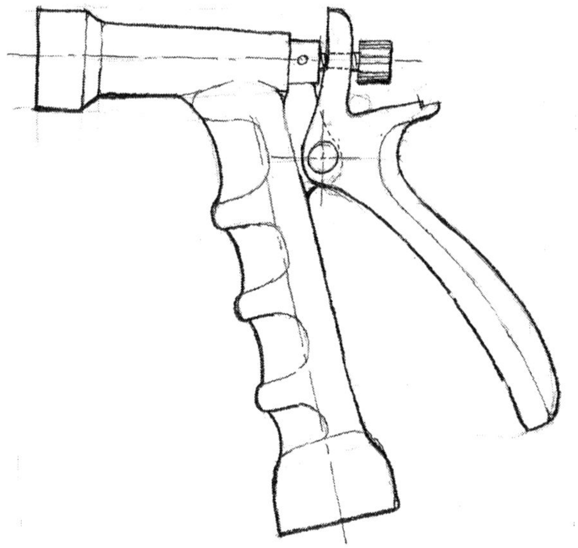

Fig. 2-6. Sketch lines can help visualize space.

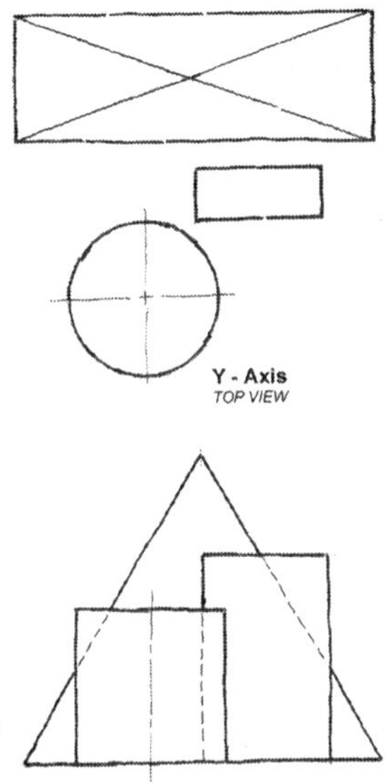

Fig. 2-5. Reading orthographic views aids in visualization.

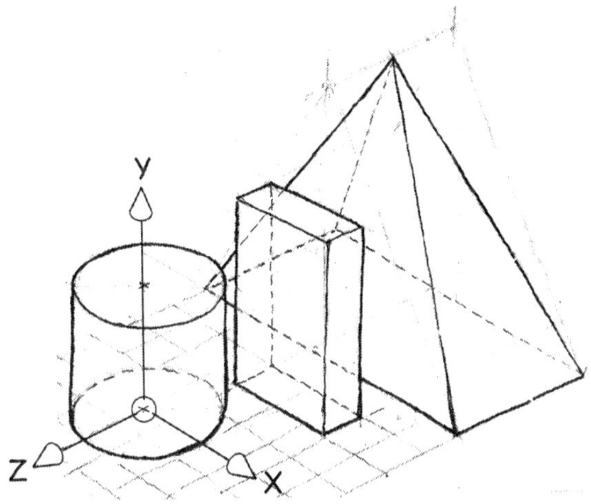

Fig. 2-7. Use every spatial cue to represent space.

Isometric in a Perspective World

Technical pictorial sketches are most easily done on an isometric construction plane (Fig. 2-8). *Iso* (equal) *metric* (measure) is a special orthographic view that places the construction plane at equal angles to the world X, Y, and Z axes. This is a special axonometric construction plane and results in several desirable conditions:

- Equal measurements can be made in any direction parallel to the three axes.
- Isometric planes—top, front, rear, bottom, and sides— are the same shape, only rotated.
- Circles in isometric planes are ellipses of the same shape, only rotated.

Isometric pictorials do not exhibit the convergence of parallel features you are used to seeing in the natural world. This is bad, because isometrics of large objects often appear unnatural. The lack of convergence is good because measurements in axis directions can be directly compared, something possible, but much more technically difficult, in perspective.

For example, chairs are shown in Fig. 2-9, both in isometric and perspective. In isometric, the heights of the chairs can be directly compared; in perspective, without additional information, it is very difficult.

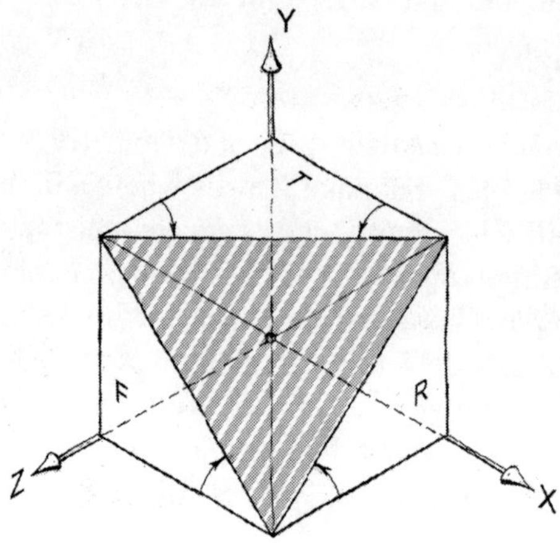

Fig. 2-8. An isometric construction plane forms equal angles with the three world axes.

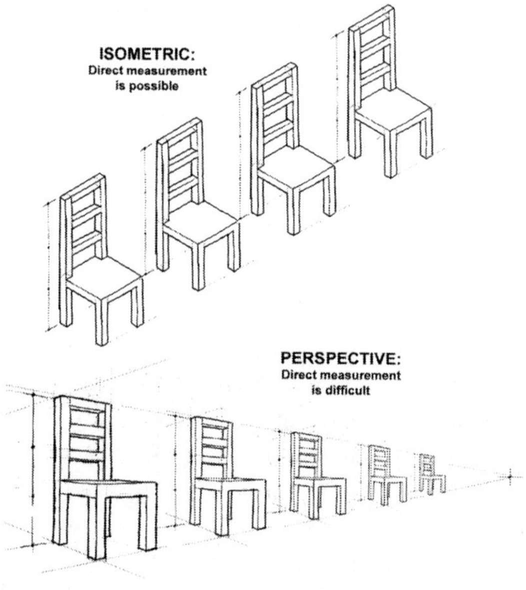

Fig. 2-9. Measurements comparable in isometric are difficult in perspective.

Sketching What You See

The first step in becoming a good visualizer is to accurately and effectively sketch what you see. Seeing in and of itself is not visualizing. Translating what you see into a sketch is a form of visualizing—as is verbally describing what you see and forming a "mental picture" of spatial relationships. Sketching what you see requires making decisions as to how a subject should be oriented in space.

First, decide on an orientation that places the prominent shape, profile, or feature to the front, near the viewer. If the object is normally encountered in a particular position or orientation, sketch it in that position. If it operates or functions in a particular position, sketch it in that orientation. Look at the product in Fig. 2-10, sketched over an isometric grid. Several positions have been explored before settling on the final orientation: the one that best shows the product's shape and features and its typical orientation of use.

After deciding on an effective orientation, sketching what you see also requires an accurate representation of size. The product in Fig. 2-10 is essentially a cylinder so the position of the part's central axis controls the view. Note that the *minor axis* of each ellipse is parallel to and coincidental with this central axis.

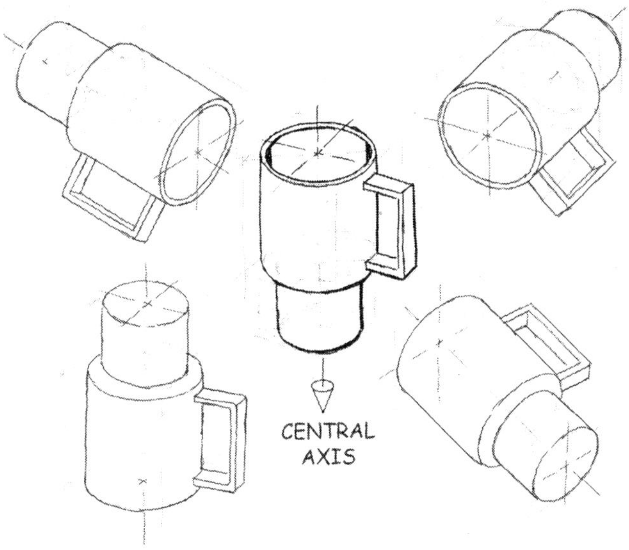

Fig. 2-10. Visualizing an effective spatial position helps make an effective sketch.

Showing Motion, Scale, and Identity

Many times a sketch uses devices other than just geometry and spatial position to represent motion, scale, and identity. Figure 2-11 shows a lexicon of sketching devices that can represent these parameters.

Movement parallel to an axis (translation) can be shown with an arrow lying in a plane containing that axis. Rotation about an axis can be shown by an arrow that wraps around the axis. Increased scale can be shown by an arrow that gets larger; decreased scale can be shown by an arrow that gets smaller. Identity can be shown with a textual call out or a numbered balloon that relates to a part name. Several additional devices have been shown in Fig. 2-11. Can you think of others?

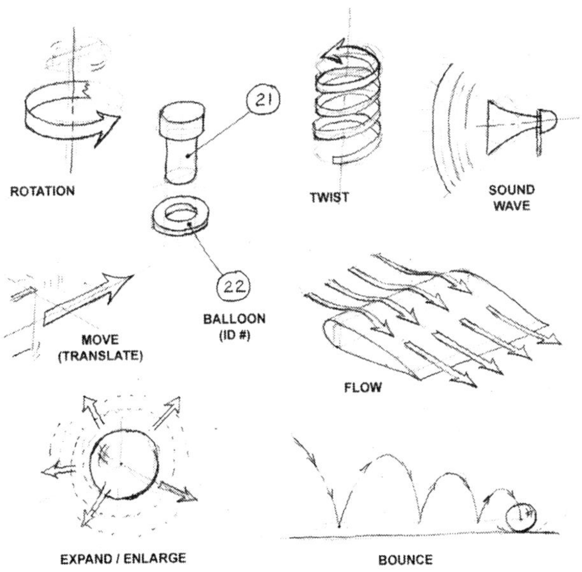

Fig. 2-11. A lexicon of sketching devices.

Practical Example

A practical example you are probably familiar with is the installation of batteries in a flashlight (Fig. 2-12, following page). In order to plan a technical animation, the sequence, timing, and spatial relationships need to be established *before* modeling and animating. Sketches of this type are called *animation storyboards* and combine many of the sketching devices we have discussed, along with textual descriptions, camera assignments, and timing.

FLASHLIGHT: Storyboard Planning Sketches for Operation Sequence.

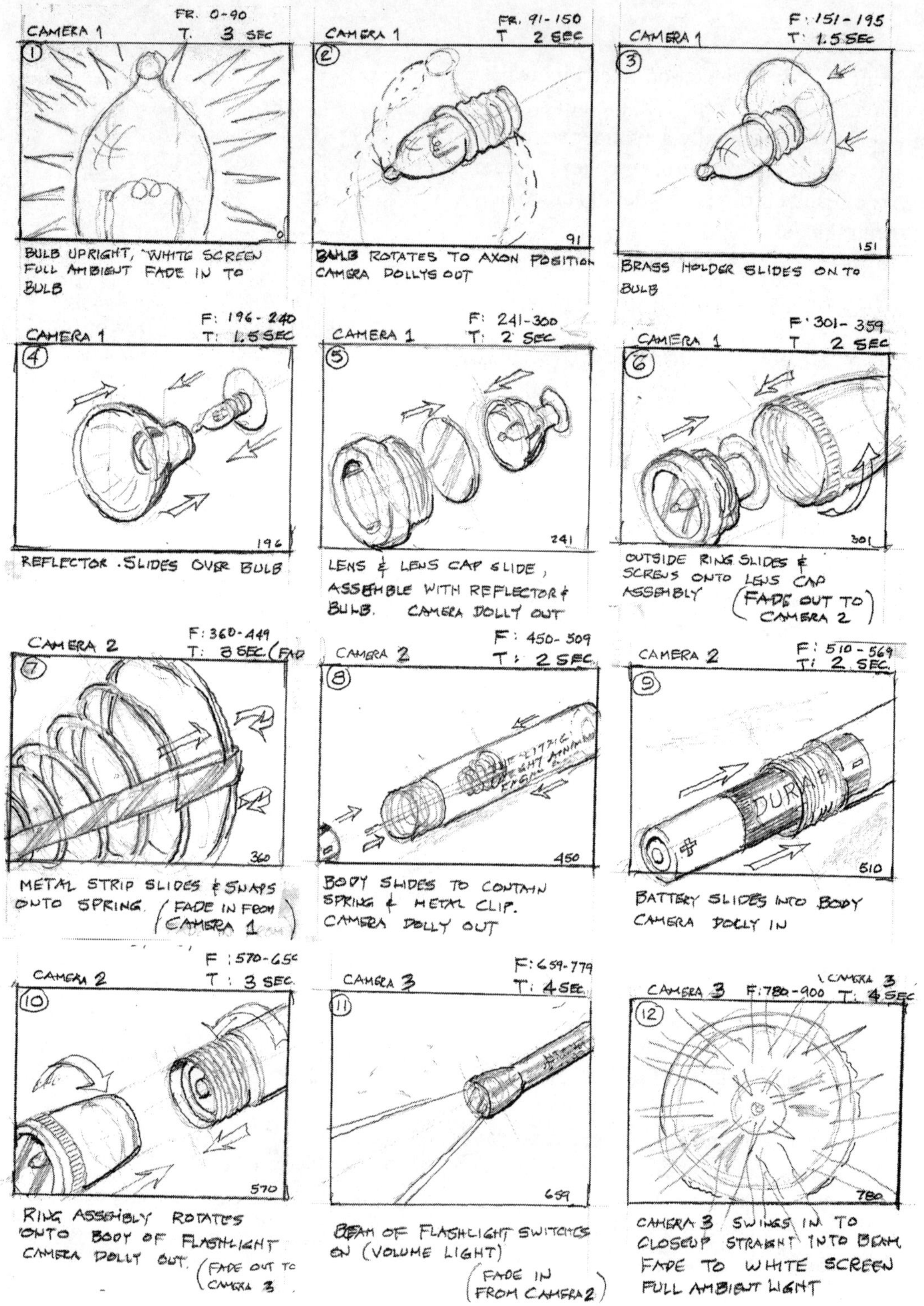

Fig. 2-12. Cues are employed for visualization of an animation sequence.

SKETCHING WHAT YOU SEE: Shown below are photographs of a series of manufactured products. Study each of the products and, based on its primary function or use, choose a logical orientation and make a sketch or series of sketches that describe it. Use isometric or orthographics grid paper as required and establish a basic scale for each sketch.

Problem 2.1 a. BINOCULARS

Problem 2.1 b. CAULK GUN

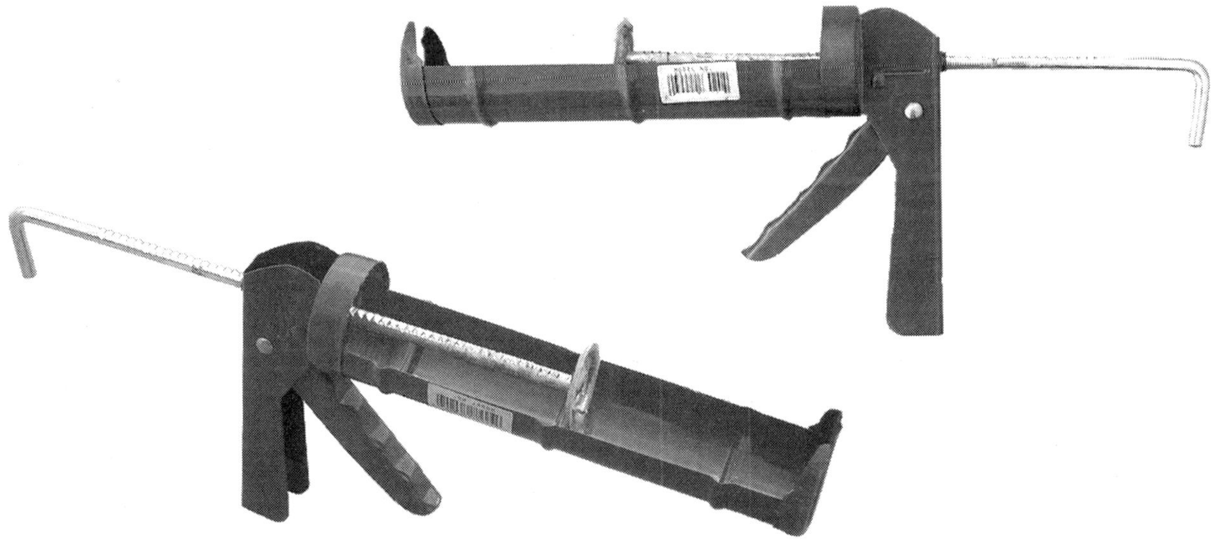

Drawn By:		FREEHAND SKETCHING	
	Scale	TITLE SKETCHING WHAT YOU SEE	
Date	Sheet	Drawing Number A	2.1

SKETCHING WHAT YOU SEE: Shown below are photographs of a series of manufactured products. Study each of the products and, based on its primary function or use, choose a logical orientation and make a sketch or series of sketches that describe it. Use isometric or orthographics grid paper as required and establish a basic scale for each sketch.

Problem 2.2 a. CORDLESS DRILL

Problem 2.2 b. TOY FIRE TRUCK

Drawn By:		FREEHAND SKETCHING	
	Scale	TITLE SKETCHING WHAT YOU SEE	
Date	Sheet	Drawing Number A	2.2

SKETCHING WHAT YOU SEE: Shown below are photographs of a series of manufactured products. Study each of the products and, based on its primary function or use, choose a logical orientation and make a sketch or series of sketches that describe it. Use isometric or orthographics grid paper as required and establish a basic scale for each sketch.

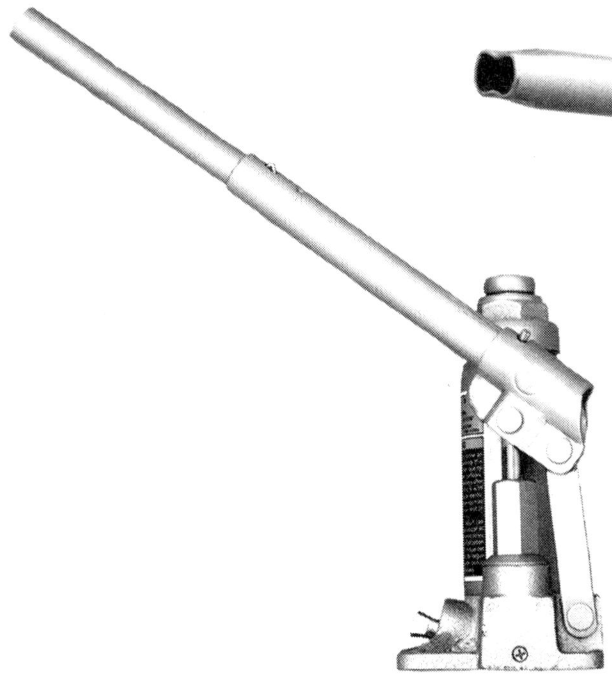

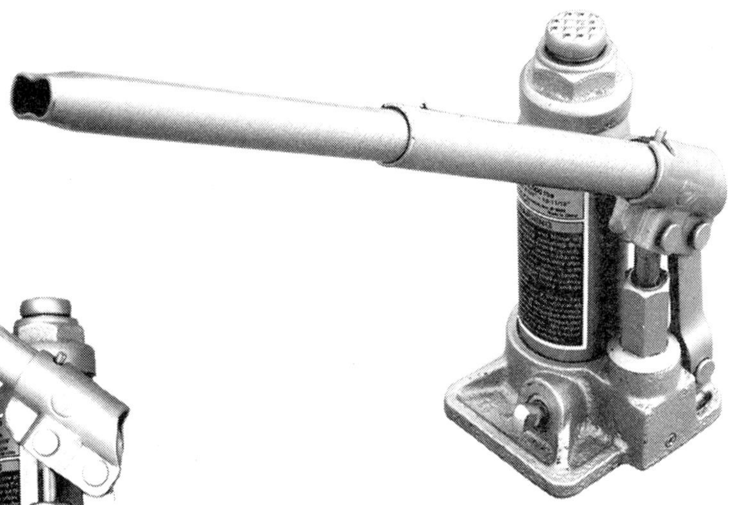

Problem 2.3 a. HYDRAULIC JACK

Problem 2.3 b. SPRAYER NOZZLE

Drawn By:		FREEHAND SKETCHING	
	Scale	TITLE SKETCHING WHAT YOU SEE	
Date	Sheet	Drawing Number A	2.3

3

3D Modeling

Coordinated with VMGED Chapter 6

The Importance of Sketching

The importance of freehand technical sketching for the purpose of graphically conceiving and communicating preliminary design ideas cannot be overemphasized. The process of sketching promotes spatial thinking, visualization, and logical, systematic ways of creating geometry. Computer-aided design tools, particularly 3D modeling software, may then be efficiently used to construct the engineering geometry for design, analysis, manufacturing, and documentation.

Although technical sketching may appear artistic, it is not considered artistic drawing. It should be thought of as a convenient method for quickly expressing and communicating ideas graphically. Freehand technical sketching is highly portable in that it requires only pencil and paper. Although any type of paper may be used for sketching, square and isometric grid papers are very popular because the grid lines make sketching and measurement easier. Figure 3-1 shows typical use of square and isometric grid paper for technical sketching. Throughout this book, problems are presented on square and isometric grid paper.

Fig. 3-1. Sketching on square and isometric grids.

Sketching Technique

Technical sketching is done freehand and rapidly; not with a straight edge and drawing instruments. Sketches should be made with dark, clearly visible lines that will be easily interpreted. For the purposes of this text, we will assume that all sketches will be made with either a wooden or mechanical graphite pencil, and not with ink.

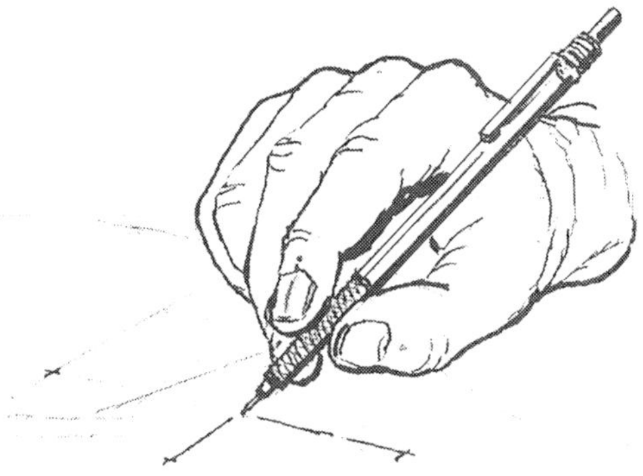

Fig. 3-2. Hand position for technical sketching.

When sketching, rest the pencil on the second finger and hold it by the thumb and index finger as shown in Fig. 3-2. In order to avoid tension or fatigue, allow the forearm to rest on the table or drawing surface while the wrist and fingers are used to sketch. When creating a sketch, do not tape the paper down. Insead, leave it free so it can be turned to a convenient angle for sketching.

Straight lines are normally sketched by first marking the endpoints of the line. This provides a starting point and an end point for each line. Horizontal lines, as shown in Fig. 3-3(a), are typically sketched from left to right with an easy motion of the wrist and fingers pivoted about the muscle of the forearm. Left-handed people normally sketch horizontal lines from right to left. Vertical lines, as shown in Fig. 3-3(b), are best sketched with a downward, pulling motion. As you sketch each line, it is a good idea to keep your eye on the end point of the line.

To sketch curved lines for arcs, cylinders, and rounded shapes, first draw lines to indicate the center of the curve. Next, insert points along the path of the curve as shown in Fig. 3-4. For a cylindrical shape, between four and eight points are recommended. Now, complete the curved arc with a light construction arc and check to see that the curve is properly proportioned. Finally, finish the curve with a bold, dark line.

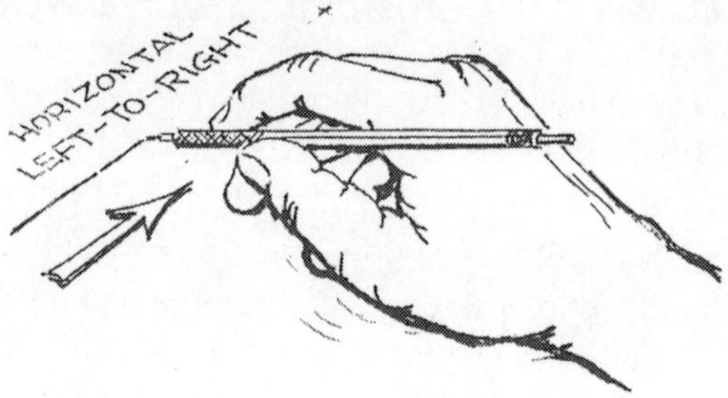

Fig. 3-3(a). Sketching horizontal lines.

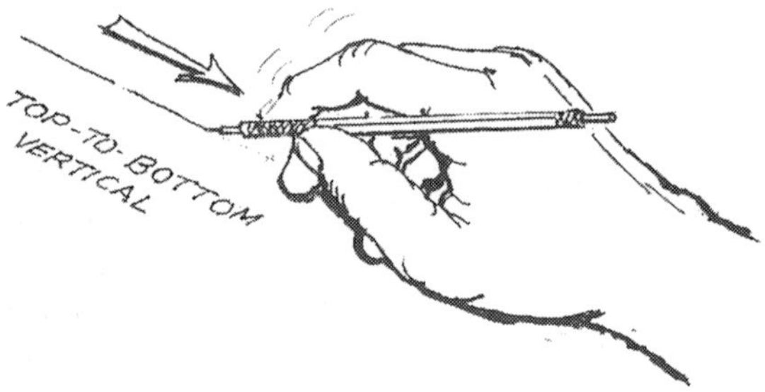

Fig. 3-3(b). Sketching vertical lines.

Model Building—General Conditions

Computer-aided design (CAD) is used to generate views for engineering drawings by one of two methods. Views may either be “electronically drafted” or constructed in 2D CAD, as shown in Fig. 3-5(a), or projected and “extracted” from revolved three-dimensional models (or views) in 3D CAD, as shown in Fig. 3-5(b).

2D CAD: Producing two-dimensional orthographic multiview drawings with 2D CAD has many similarities to sketching and manual drafting. Views are created and aligned using 2D geometric construction methods such as template or construction lines, tangency, polygons, trimming and extending of lines, automatic line types, and a variety of other tools. Construction of isometric and auxiliary views in 2D CAD also relies on similar 2D construction methods such as boxing out, grid construction, and normal projection from lines representing the edge views of surfaces.

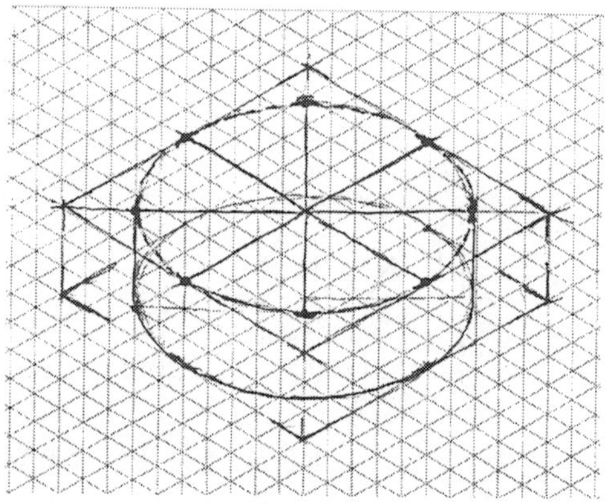

Fig. 3-4. Laying out points to sketch a curve.

3D CAD: Producing orthographic multiview drawings using 3D CAD is more a matter of moving the viewer or rotating the object(s) so that the desired view is obtained, as shown in Fig. 3-5(b). Standard views of the model—such as front, top, or right side—are typically aligned with the CAD program's 3D axis system. Since any view is possible with a three-dimensional model, isometric pictorial and auxiliary views only require revolving the model or the user's line of sight into the correct viewing position. In auxiliary views, this is accomplished by aligning the viewer's line of sight at a right angle (normal) to a particular inclined or oblique feature on the model by means of a movable construction plane with its own local coordinate system. If actual flat, two-dimensional views are required, the 3D CAD program must also have the capability to convert or flatten the three-dimensional model into a true two-dimensional representation on any user-defined construction plane containing a desired view, as shown in Fig. 3-6. The resulting 2D view may then be used for drafting and documenting purposes.

Although faster and more accurate than manual drawing methods, 2D CAD has the same limitation as the drawing board: It is two-dimensional. As a tool for visualizing and creating objects that may be viewed from their front, top, side, auxiliary, or isometric positions, 3D CAD offers much more power than 2D CAD. Therefore, the focus of this chapter, and the premise of this book, is on methods of using 3D CAD to visualize, model, and generate required views for engineering drawings directly from 3D models.

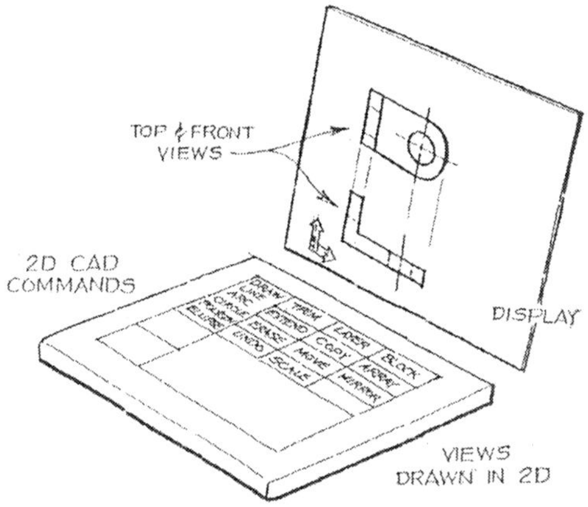

Fig. 3-5(a). Views constructed in 2D CAD.

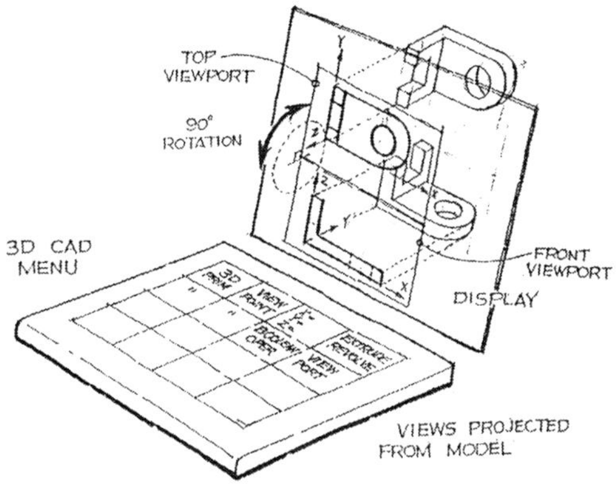

Fig. 3-5(b). Views projected from a 3D CAD model.

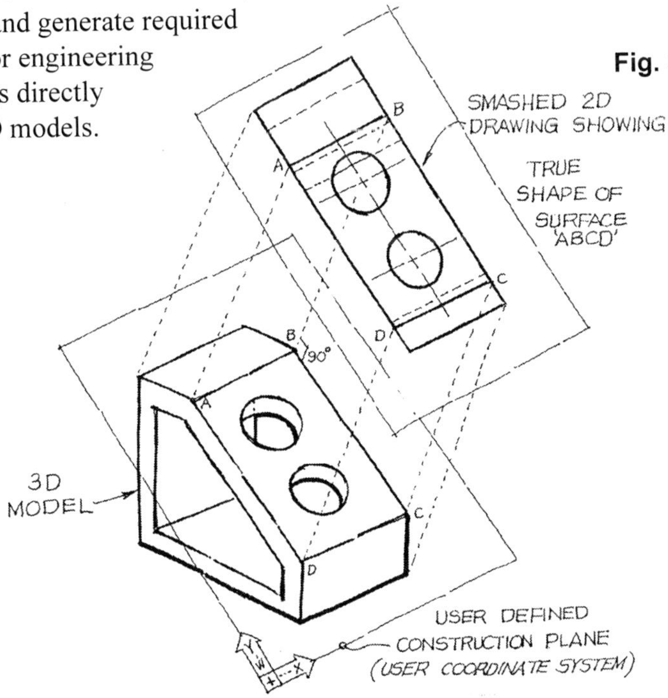

Fig. 3-6. Smashing 2D images from 3D CAD models.

Movable Construction Planes in 3D

All 3D CAD programs provide the user with virtual "3D space" in which to construct models. In order to use this modeling space accurately and efficiently, the user must have some method of controlling the position and size of every 3D object in that space. Hence, 3D CAD programs furnish the user with a predetermined and fixed Cartesian coordinate system containing X, Y, and Z axes for referencing the absolute position of each vertex, edge, surface, and object in the model. This coordinate system is commonly referred to as the *world coordinate system.*

In the real world, measurements are seldom made from a single reference point that is absolute or fixed. Instead, measurements are taken relatively or locally from any convenient point to any other point. For example, if the corner of a room is considered to be the origin of the world coordinate system (0, 0, 0), how would you measure the width of a table inside the room? Using the world coordinate system, you must locate each corner of the table relative to the world origin, i.e., (4, 6, 3) for the first table corner and (5, 9, 3) for the opposite diagonal. So, how wide is the table? A rather awkward way of measuring, don't you think? To solve this problem, a movable local coordinate system—called the *user coordinate system* (UCS)—is supplied. Figure 3-7 illustrates this concept. The local UCS can be repositioned on-the-fly, similar to a tape measure, and placed anywhere and at any orientation within the modeling space.

One of the most powerful ways of using a movable user coordinate system is to employ it as a reference or construction plane. A *construction plane* (also referred to as a work plane or a reference plane) can be thought of as an infinitely large 2D drawing plane that can be positioned anywhere in 3D space. Since most people feel comfortable drawing in 2D, the construction plane can be used as a drawing surface as well as a plane on which to position 3D objects. Unlike 2D CAD, the construction plane can be moved freely to any other location and orientation in 3D space. In most popular CAD programs the construction plane, and its local coordinate system, is visually indicated by the UCS icon, showing the orientation of the local axes—$X_L Y_L Z_L$. Figure 3-7 illustrates the concept of the construction plane shown with a UCS icon.

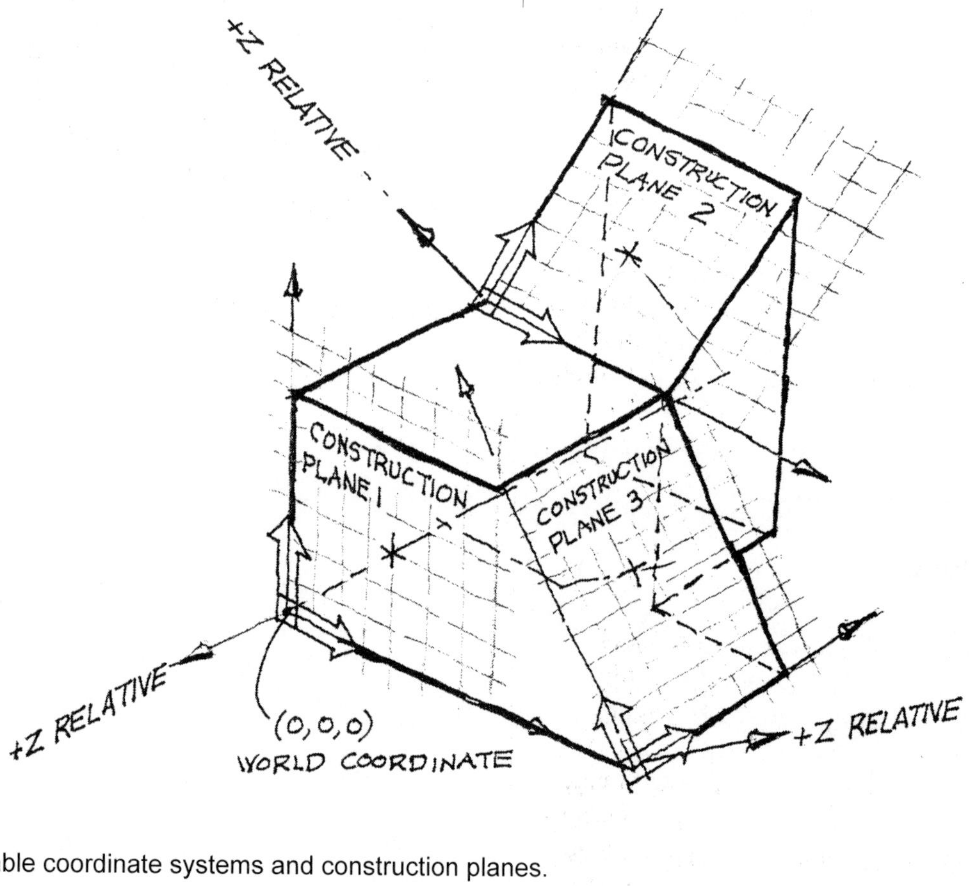

Fig. 3-7. Movable coordinate systems and construction planes.

Modeling Primitives

Perhaps the most common method of three-dimensional modeling is by combining 3D geometric *primitives*. Most 3D CAD programs come with a standard set of primitives, which include, but are not limited to, the box (prism), cylinder, cone, and sphere. The primitives may then be selectively combined to build the various features of an object, as shown in Fig. 3-8. Primitives are typically specified by the user according to their overall dimensions, such as diameter and height for a cylinder. When primitives are input by the user, they are typically attached to or positioned with respect to the coordinates of the active construction plane.

There are three different representational methods used for geometric modeling in 3D CAD programs: wireframe, surface, and solid models. Of these three methods, solid primitives give the most complete and unambiguous description of an object. In other words, a hole drilled through a block will be treated as a true hole in a block (solid method) and not as a circular opening in a hollow box (surface method), or a circle floating on an open plane (wireframe). Because the primary focus of this text is on sketching and describing "real world" objects made of solid materials such as metal, plastic, or wood, all objects in this chapter and throughout the book are referred to and treated as solid objects.

In 3D CAD, not all objects have sharp, well-defined edges that can be shown easily with lines. To describe the boundaries of surfaces on curved primitives such as cylinders, cones, and spheres, special lines are required. These display lines, often referred to as *tessellation lines*, are useful in visually clarifying the description of complex curved surfaces.

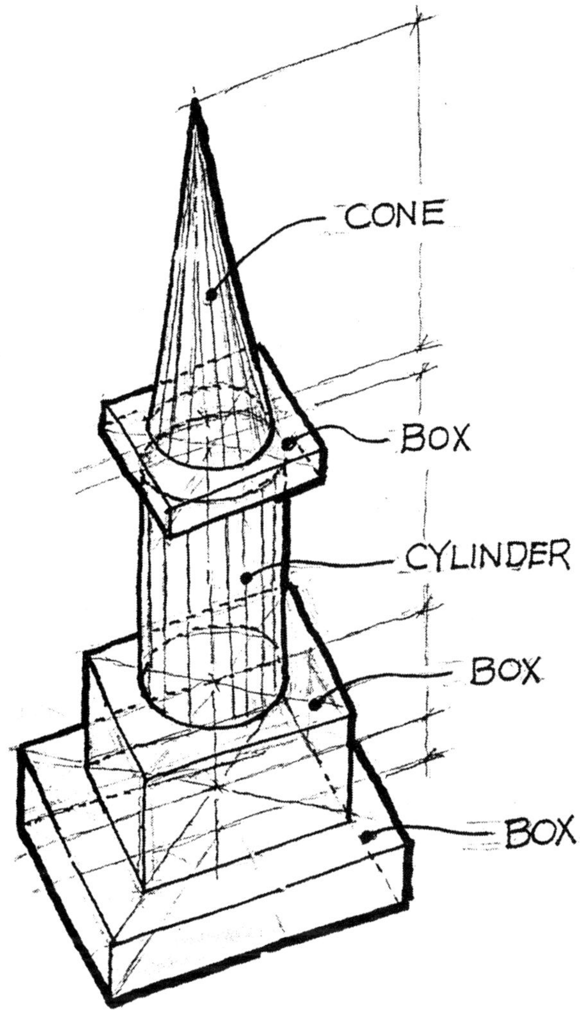

Fig. 3-8. Building with 3D primitives.

Boolean Operations

As described previously, a primary method of model building is by stacking a series of 3D primitive building blocks to "additively" construct a model. Although useful as a starting point for solid modeling in 3D CAD, this method is limited unless the user can define a range of relationships between overlapping or connected primitives. These relationships are called *Boolean operations*, named after English mathematician George Boole (1815-

1864), who developed a mathematical system in which all variables are assigned a value of one or zero. Boolean operations allow the user to perform three types of operations: union, difference, and intersection. In all Boolean operations, the area of interest is where common volumes of 3D shapes intersect or overlap. It is at the zone of overlap or intersection where the Boolean operation takes place. Although typically performed on only two objects at a time, Boolean operations can be performed on sets of three or more objects at once.

The *union operation*, shown in Fig. 3-9(a), is essentially an additive operation with all selected geometry being merged into a new composite object. Nothing is discarded and the order of operation is unimportant. The new object may be thought of as a single "higher order" primitive that retains an identifiable subset of original shapes. In other words, the user may "undo" a higher order primitive and break it down into its composite primitives if need be. This continuing or branching relationship of operations is referred to as a *Boolean tree*.

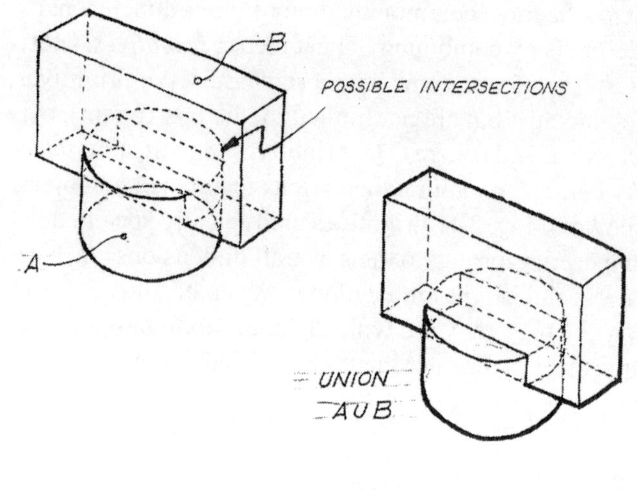

Fig. 3-9(a). The union operation.

The *difference operation* is a subtractive operation that creates a composite solid by subtracting the overlapping volume of one set of primitives from another set of primitives. Note, as shown in Fig. 3-9(b), there are two possible outcomes of the difference operation. The order determines the result. The difference operation is a useful technique for doing simple but important modeling operations such as creating holes in solids by subtracting a cylinder (conceptually, a drilling operation).

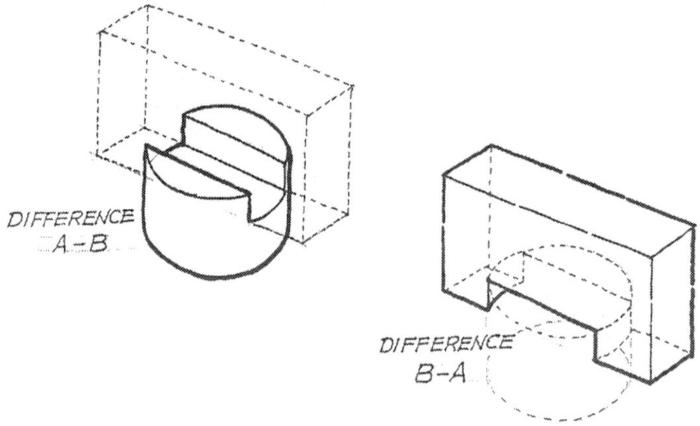

Fig. 3-(9b). The difference operation.

The *intersection operation* takes only the overlapping 3D volume common to all selected primitives and makes that shared space the resulting object. The order of operation is unimportant. Although less intuitive than union and difference operations, the intersection operation is a powerful method of creating complex objects in a single step. Figure 3-9(c) shows the finished geometry resulting from the intersection of two solid primitives.

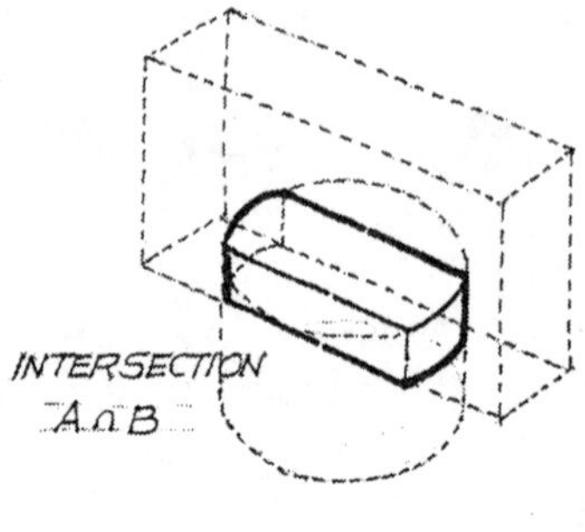

Fig. 3-(9c). The intersection operation.

Extruded and Revolved Objects

Solid objects with thin walls and complex symmetrical exterior shapes cannot always be modeled easily with geometric primitives and Boolean operations. Many parts, because of their shapes, can be created as *extruded* and *revolved* objects by sweeping their profiles along or around an axis. Just as you can mentally decompose a complex solid object into regular primitives and Boolean operations, you can imagine the cross-section through a solid object and create an accurate profile of that shape. The profile can then be used in one of two ways to create a solid object.

Extruded solid objects are typically made by sweeping a closed profile along an axis. Normally, the profile is perpendicular to the axis of extrusion, as shown in Fig. 3-10(a).

Revolved solid objects are made by sweeping or revolving a profile shape about an axis. Although solids of revolution are usually swept through 360 degrees of circular rotation, any convenient angle can be used, as shown in Fig. 3-10(b). Compare the different results of the same profile being extruded in (a) and revolved in (b).

Modeling by Sketching Exercises

The sketching exercises on pages 47-57 are designed to provide practice in the model building techniques commonly used in 3D CAD and solid modeling. The logic involved in these exercises is intended to serve as a useful starting point for visualizing, sketching, and modeling the problems in this text that follow.

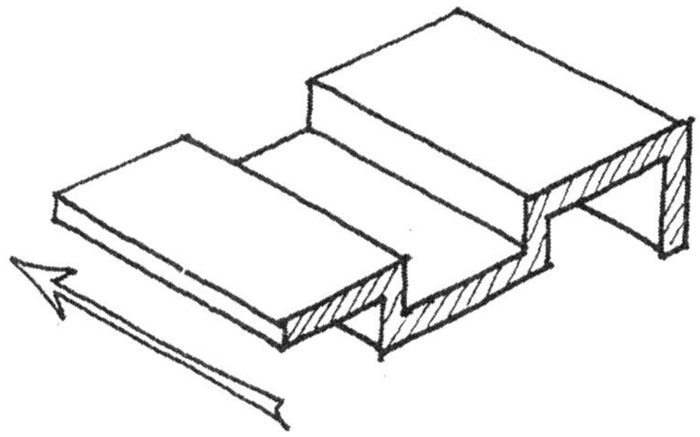

Fig. 3-10(a). Example of extruded shape.

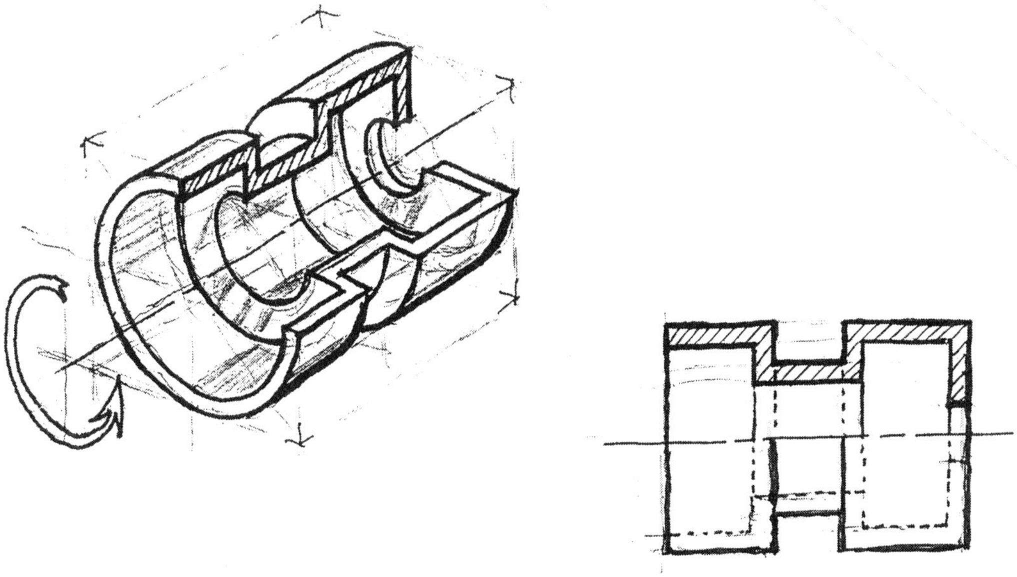

Fig. 3-10(b). Example of revolved shape.

GIVEN THE SIZE DESCRIPTIONS BELOW, MAKE AN ISOMETRIC SKETCH OF EACH PRIMITIVE BASED ON THE ORIENTATION OF THE USER COORDINATE SYSTEM (UCS). NOTE THE + AND – DIRECTION OF THE Z-AXIS. DARKEN IN ONLY VISIBLE EDGES.

SKETCH A CONE CENTERED AT (X=0, Y=0, Z=0). DIAMETER=6 UNITS, ALTITUDE = 8 UNITS ON +Z-AXIS.

EXAMPLE

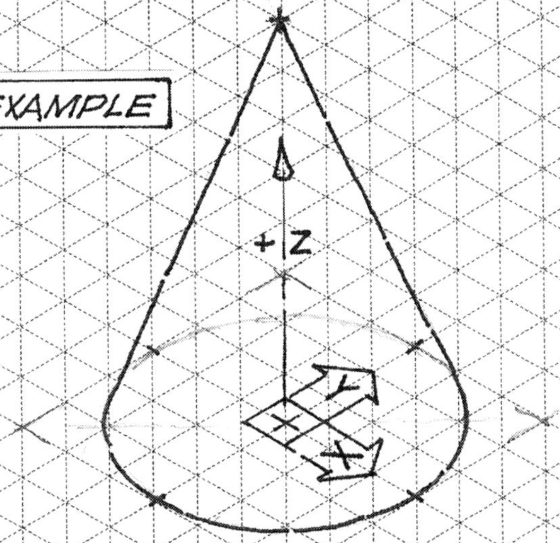

SKETCH A CYLINDER CENTERED AT (0,0,0). DIAMETER= 6 UNITS, LENGTH= 8 UNITS ON –Z-AXIS.

(1)

SKETCH A BOX WITH BASE CORNERS AT (0,0) AND (8,5). HEIGHT = 4 UNITS ON +Z-AXIS.

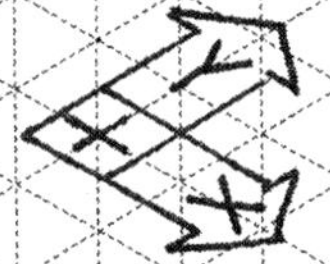

(2)

SKETCH A CONE CENTERED AT (0,0,0). DIAMETER = 6 UNITS, ALTITUDE = 9 UNITS ON -Z-AXIS.

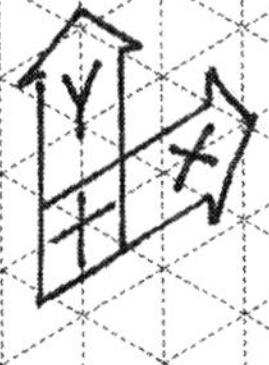

(3)

Drawn By:		FREEHAND SKETCHING	
	Scale	TITLE PRIMITIVE ORIENTATION	
Date	Sheet of	Drawing Number A	3.1

GIVEN THE TOP & FRONT VIEWS OF THE CITY PRIMITIVES, MAKE A PICTORIAL SKETCH OF THE SCENE ON A SEPARATE SHEET OF ISOMETRIC GRID PAPER. DARKEN IN ONLY VISIBLE EDGES. OPTIONAL: SHADE IN ALL SURFACES FACING TO THE RIGHT.

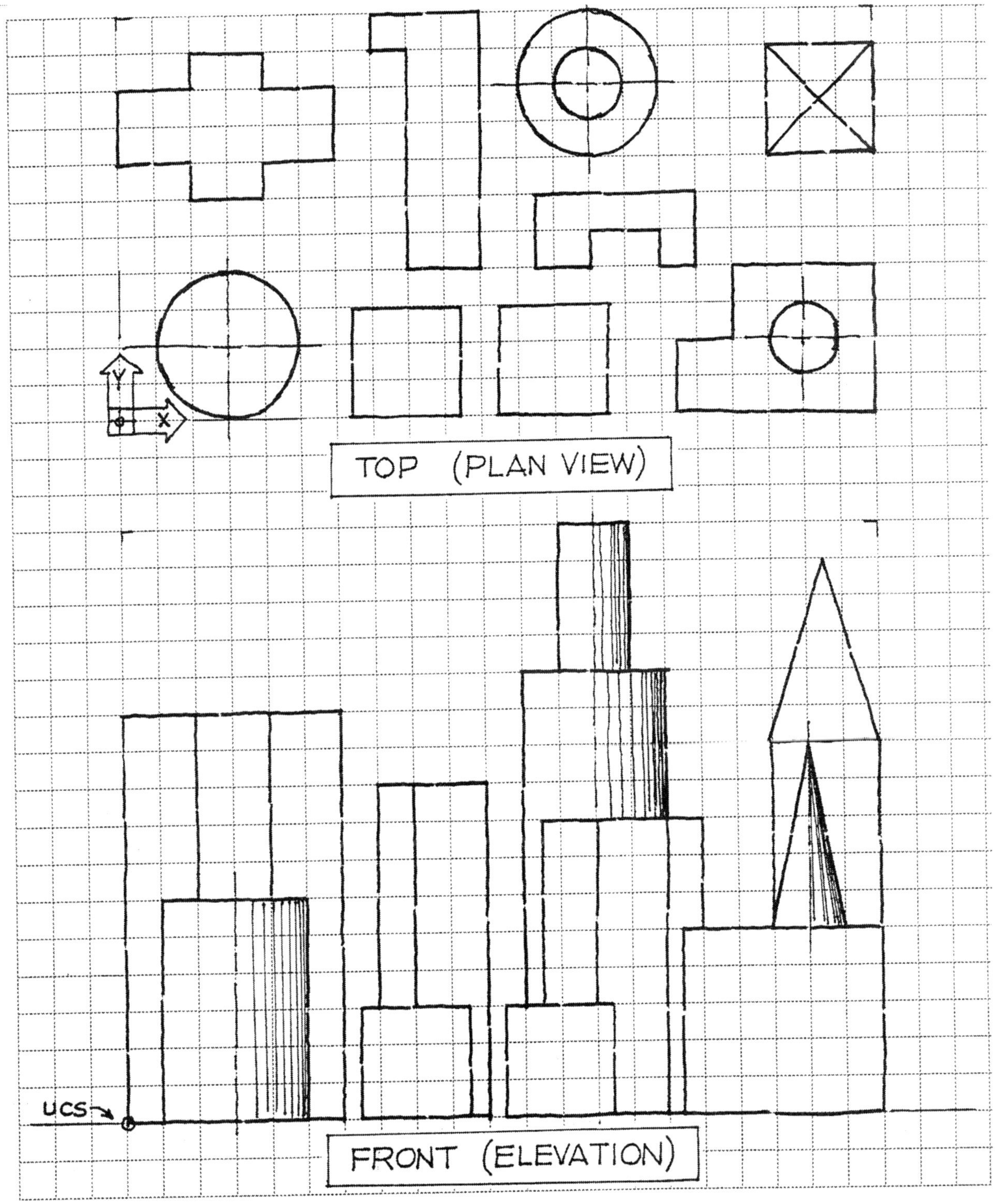

Drawn By:		FREEHAND SKETCHING	
	Scale	TITLE PRIMITIVE VISIBILITY	
Date	Sheet of	Drawing Number A	3.2

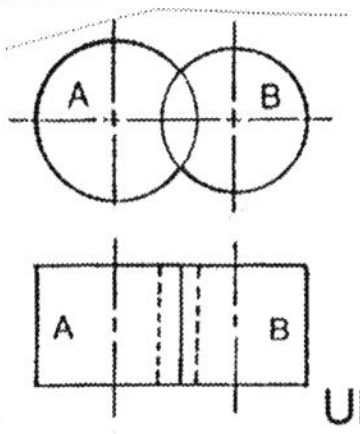

BASED ON THE OVERLAPPING SOLID PRIMITIVES SHOWN BELOW, AND THE BOOLEAN OPERATIONS ASSIGNED, DARKEN IN ALL VISIBLE EDGES OF THE RESULTING COMPOSITE SOLID AS SHOWN IN THE EXAMPLES.

UNION (A + B)	DIFFERENCE (A - B)	DIFFERENCE (B - A)	INTERSECTION (A ∩ B)

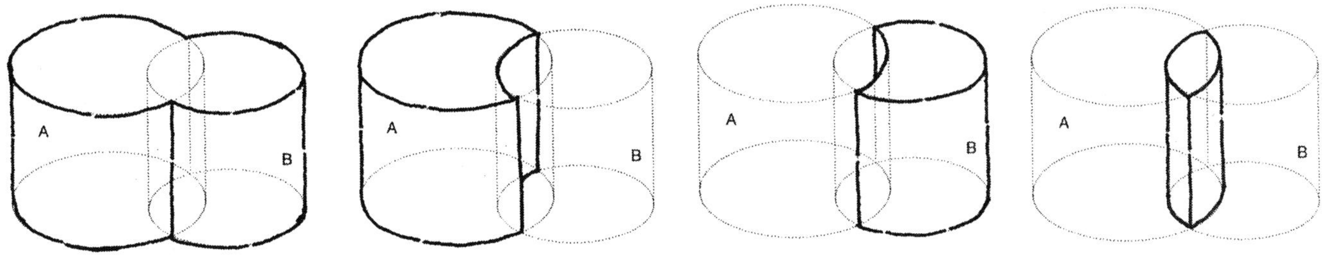

EXAMPLES I. II. III. IV.

(1) ☐ UNION (A + B) ☐ DIFFERENCE (A - B) ☐ INTERSECTION (B∩A) ☐ DIFFERENCE (B - A)

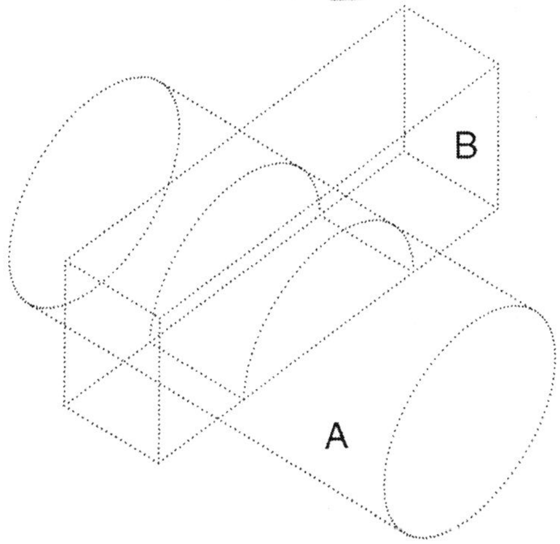

(2) ☐ UNION (A + B) ☐ DIFFERENCE (A - B) ☐ INTERSECTION (B∩A) ☐ DIFFERENCE (B - A)

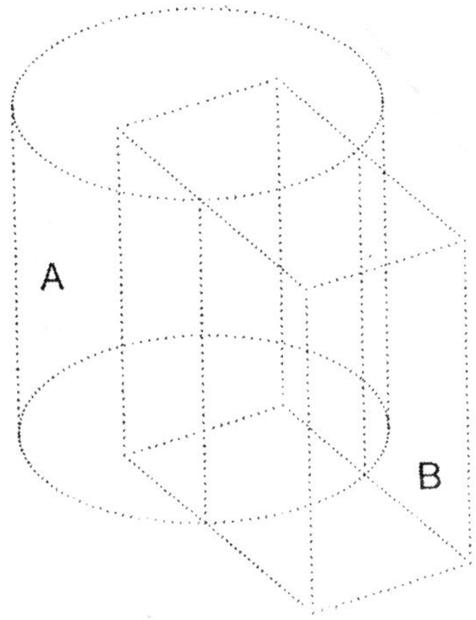

(3) ☐ UNION (A + B) ☐ DIFFERENCE (A - B) ☐ INTERSECTION (B∩A) ☐ DIFFERENCE (B - A)

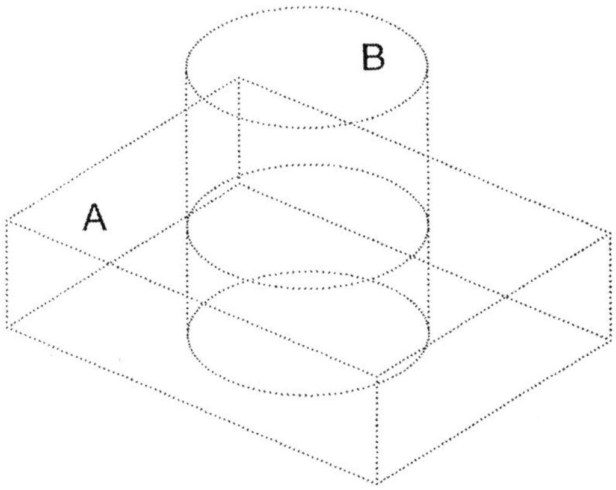

(4) ☐ UNION (A + B) ☐ DIFFERENCE (A - B) ☐ INTERSECTION (B∩A) ☐ DIFFERENCE (B - A)

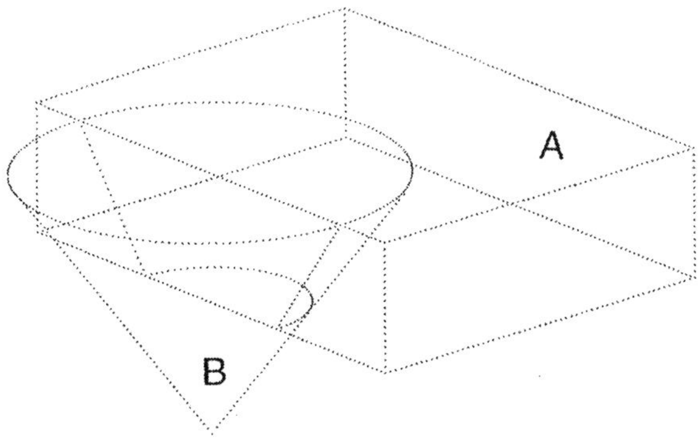

Drawn By:		FREEHAND SKETCHING FOR CAD	
	Scale	TITLE BOOLEAN SOLIDS	
Date	Sheet of	Drawing Number A	3.3

BASED ON THE THREE PRIMITIVES SHOWN BELOW, AND THE SEQUENCE OF BOOLEAN OPERATIONS ASSIGNED, DARKEN IN ALL VISIBLE EDGES OF THE RESULTING COMPOSITE SOLID.

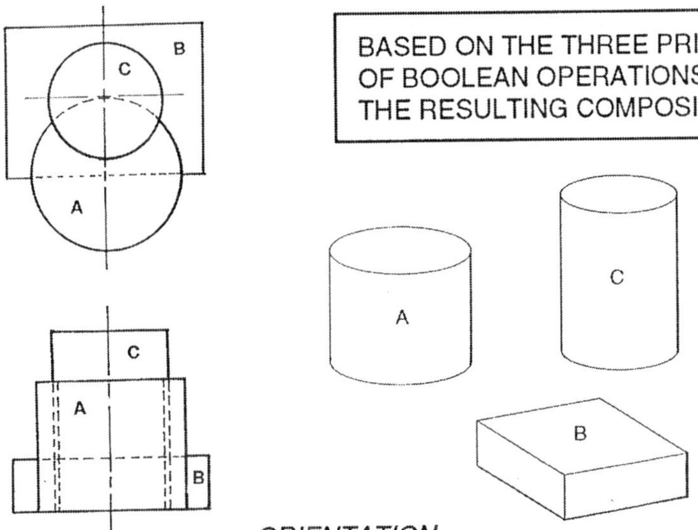

ORIENTATION

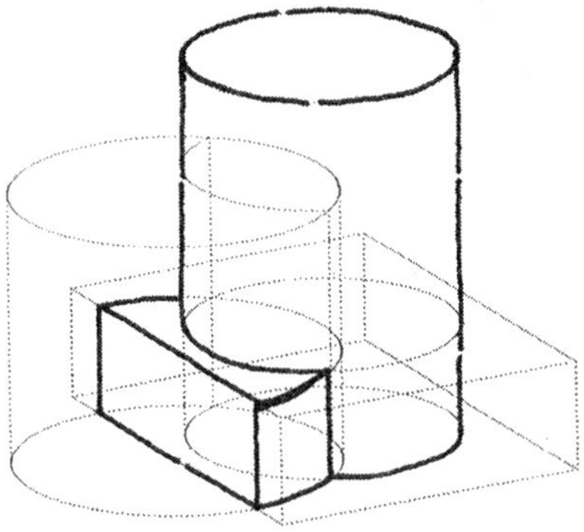

Ⓢ $(A \cap B) + C$ *EXAMPLE*

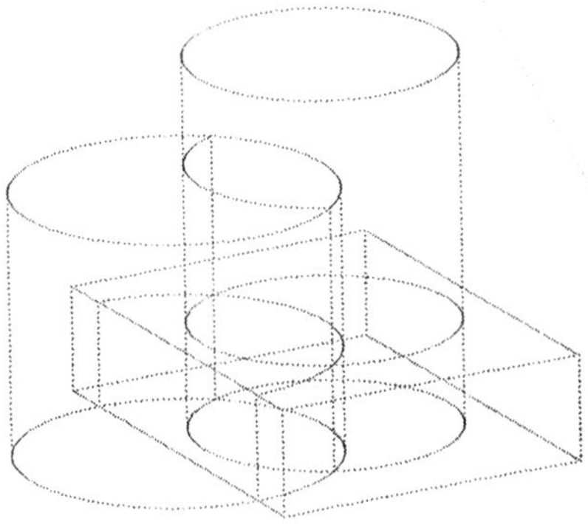

① ________

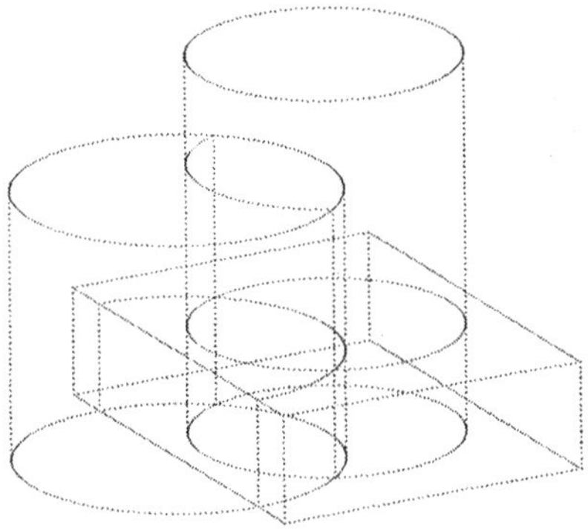

② ________

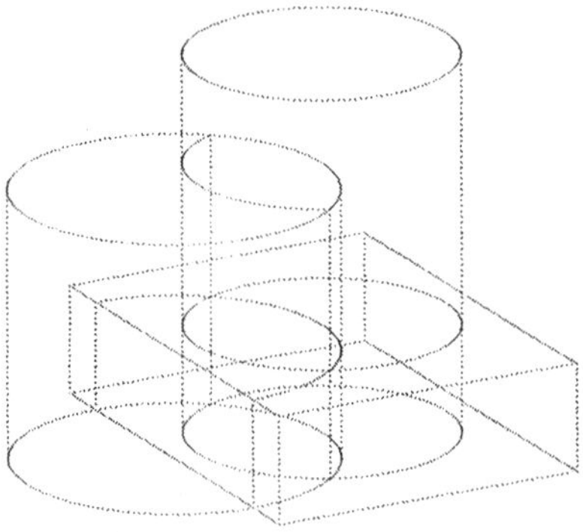

③ ________

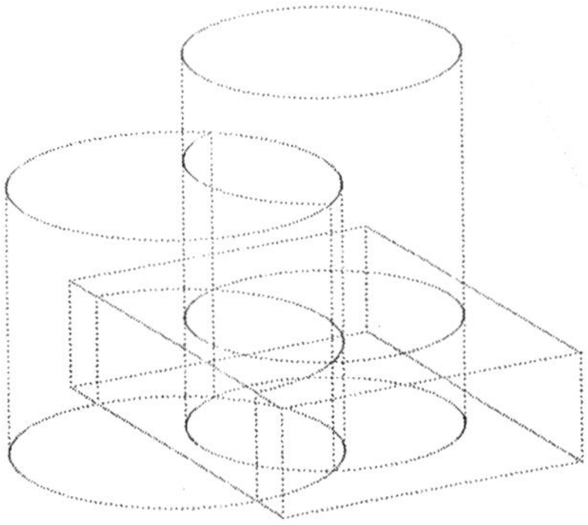

④ ________

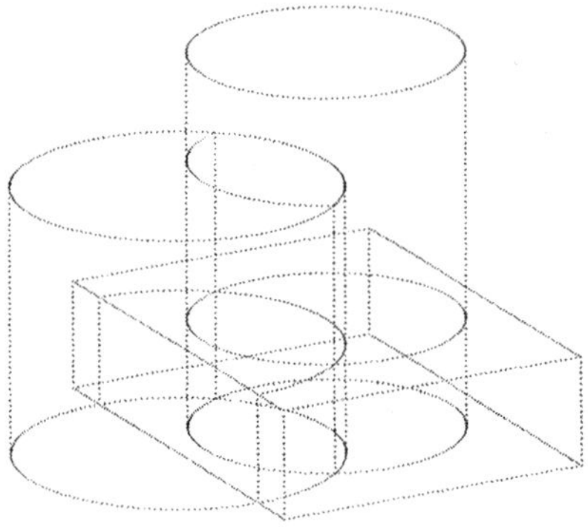

⑤ ________

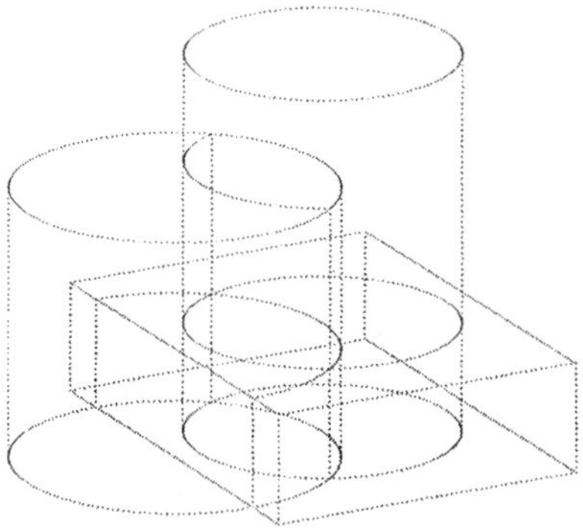

⑥ ________

Drawn By:		FREEHAND SKETCHING FOR CAD	
	Scale	TITLE BOOLEAN SOLIDS	
Date	Sheet of	Drawing Number A	3.4

EXTRUDE THE PROFILES AS SHOWN. USING THE LOGIC OF BOOLEAN INTERSECTION, DARKEN IN THE VISIBLE EDGES OF THE RESULTING SOLID MODEL.

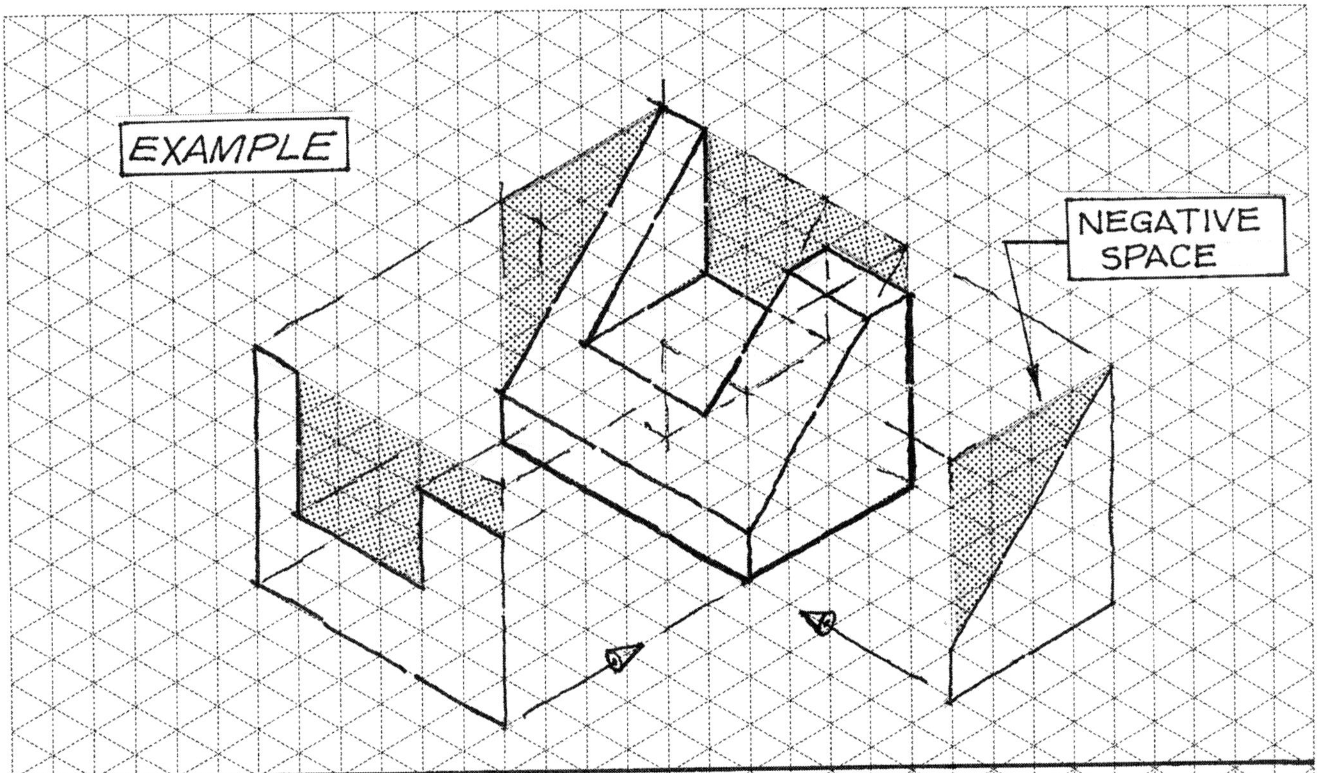

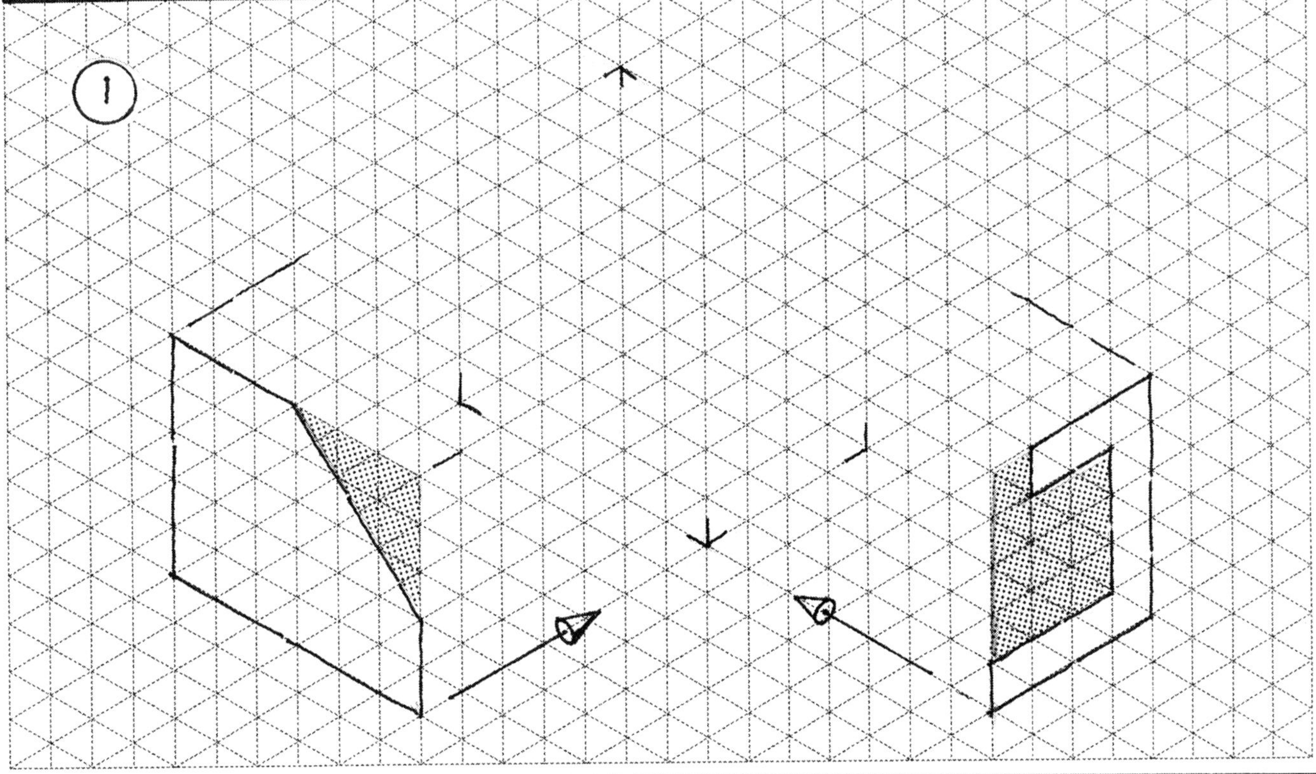

Drawn By:		FREEHAND SKETCHING FOR CAD	
	Scale	TITLE INTERSECTION	
Date	Sheet of	Drawing Number A	3.5

EXTRUDE THE PROFILES AS SHOWN. USING THE LOGIC OF BOOLEAN INTERSECTION, DARKEN IN THE VISIBLE EDGES OF THE RESULTING SOLID MODEL.

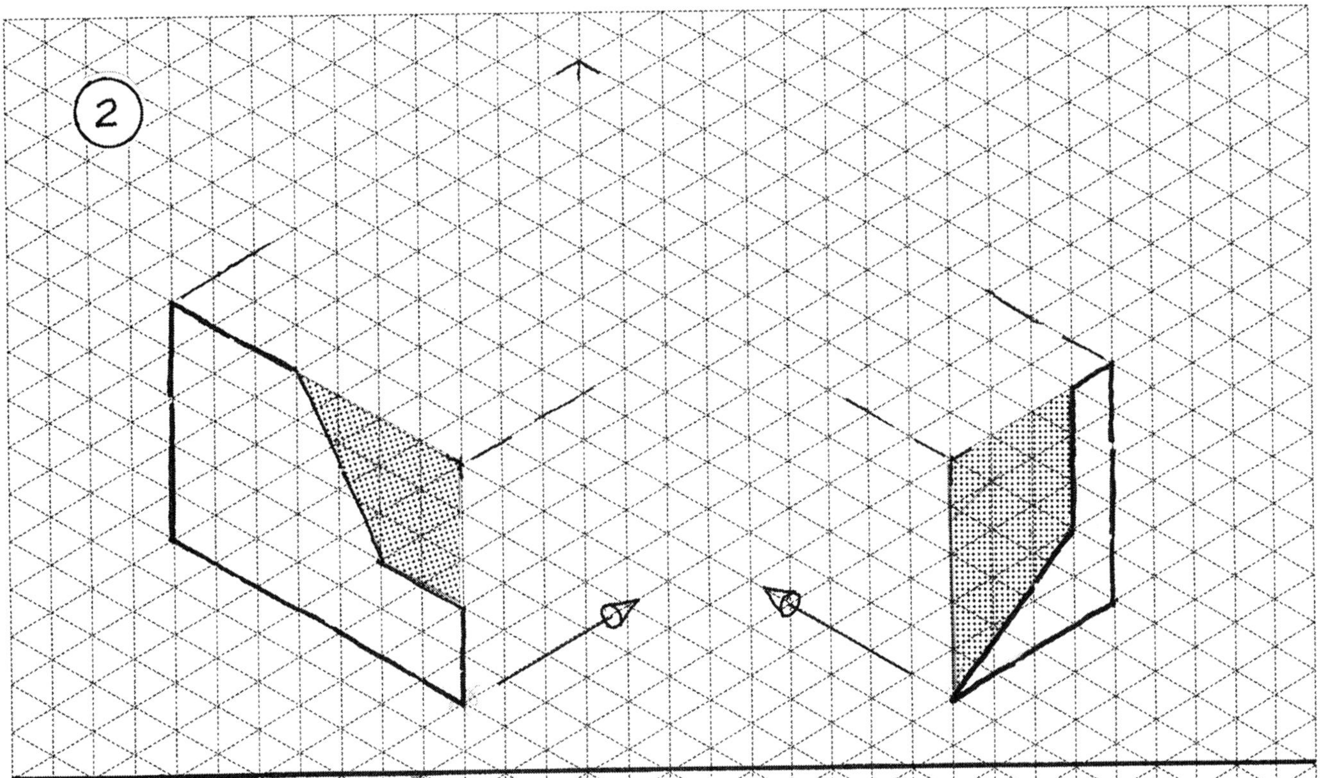

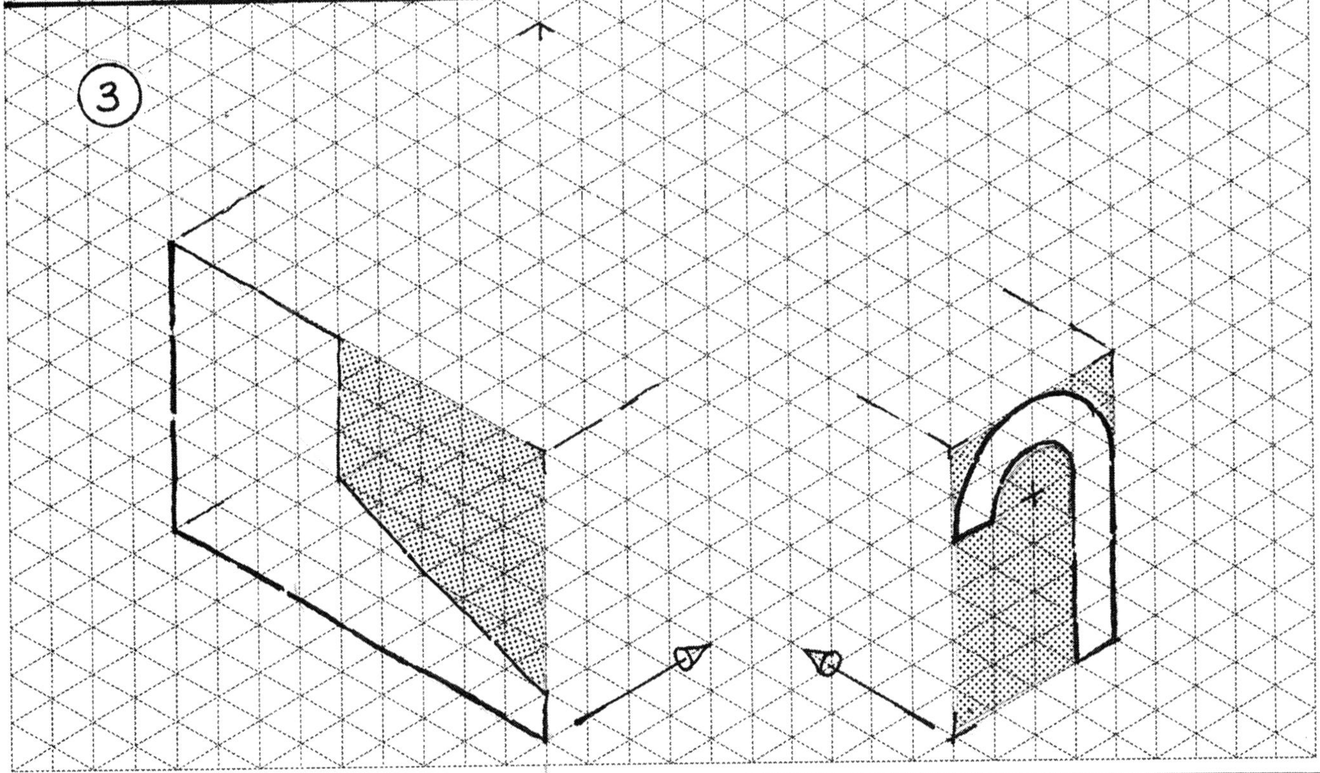

Drawn By:		FREEHAND SKETCHING FOR CAD		
	Scale	TITLE	INTERSECTION	
Date	Sheet of	Drawing Number	A	3.6

4 Orthographic Sketching

Coordinated with VMGED Chapter 10

Orthographic Views—General Conditions

An *orthographic view* (ortho, meaning at right angles) presents two of the three dimensions of height, width, and depth in one view. When these three dimensions are known, the object is unambiguous. The view can be made by different people at different locations within acceptable variation. Two or more views drawn at the same scale are usually arranged so that two share a common dimension. This allows the views to be *read,* that is, the geometry to be understood. When views are arranged in this manner they are said to be *aligned.* See Fig. 4-1. In practice, large industrial drawings may be on separate sheets, and at different scales, making reading difficult.

Orthographic views show no perspective. Parallel conditions in space are shown parallel on the orthographic drawing. The perspective in Fig. 4-1 is for comparison only and is not an aligned view.

The *front view* is usually the view that shows the characteristic shape of the object or the position in which the object is usually found. Additional views are positioned as needed above (top view), to the right or left (side views), or below (bottom view). Objects such as shims, gaskets, or plates require only one view, not a view of their thickness (Fig. 4-2). Cylindrical objects may require only two views. The more complicated the object, the more views are needed to fully describe geometry.

Sheet 4.0 on page 63 shows both pictorial and orthographic views of the same object. The *pictorial* is used for visualization. The orthographic views are used to establish the exact relationship of geometric features such as planes, holes, and the like. Viewing directions have been

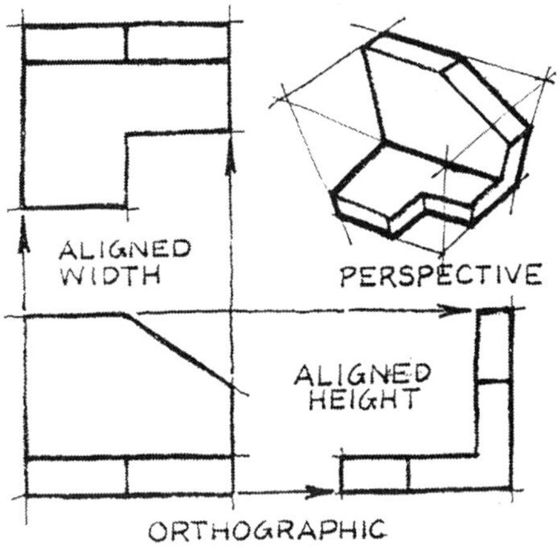

Fig. 4-1. Aligned views.

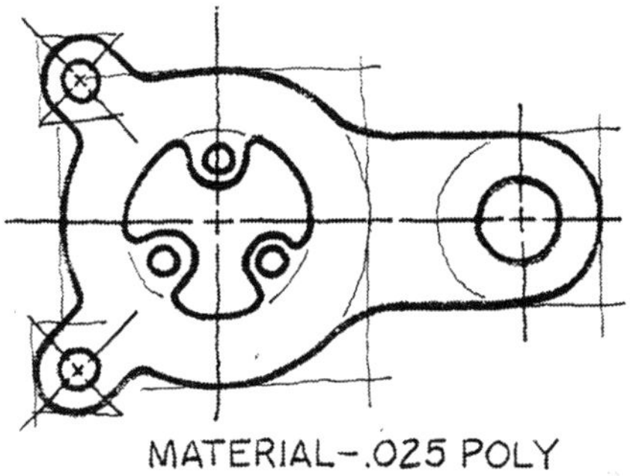

Fig. 4-2. A one-view orthographic sketch.

established in the pictorial view that correspond to the orthographic views. Study each viewing direction and the resulting orthographic view.

Representation of Intersections, Surfaces, and Limits

In orthographic views, lines are used for very specific purposes.

Solid Lines. A visible plane edge, surface intersection, or limit (Fig. 4-3).

Dashed Lines. An edge, intersection, or limit obscured by material between it and the observer (Fig. 4-4).

Lines of Symmetry. Non-geometric lines used to show the center of a hole or the axis of a cylinder (Fig. 4-5).

Assignment: On the orthographic views of sheet 4.0, identify each of the lines by type (solid, dashed, symmetry) and by geometric feature (edge, limit, intersection). Practice your technical lettering on each label.

Practical Example

In Fig. 4-6, a pictorial sketch of an electrical part has been prepared. This is often the first step in designing a part—producing a pictorial sketch. In the next chapter you will gain more experience in pictorial sketching.

Note you can visualize a part's general shape quite easily from a pictorial sketch. From the pictorial, traditional orthographic views can be prepared. Follow these steps to prepare orthographic views (Fig. 4-7):

1. Decide which views are necessary and sufficient. Thin objects only require one view and a thickness note; cylindrical objects (like Fig. 4-5) require two views. Objects with inclined or oblique surfaces may require three or more views.

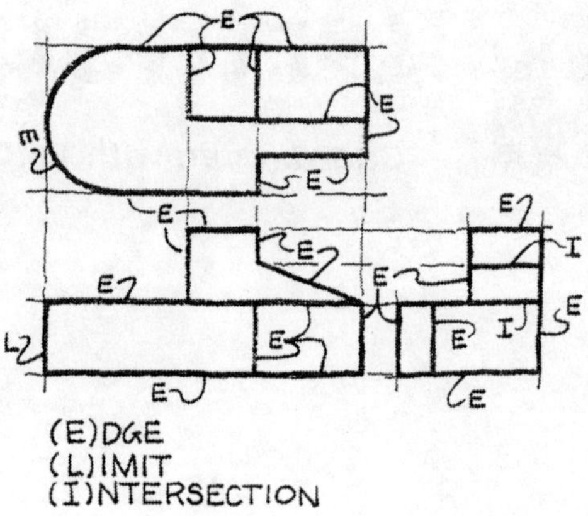

Fig. 4-3. Solid lines on a sketch.

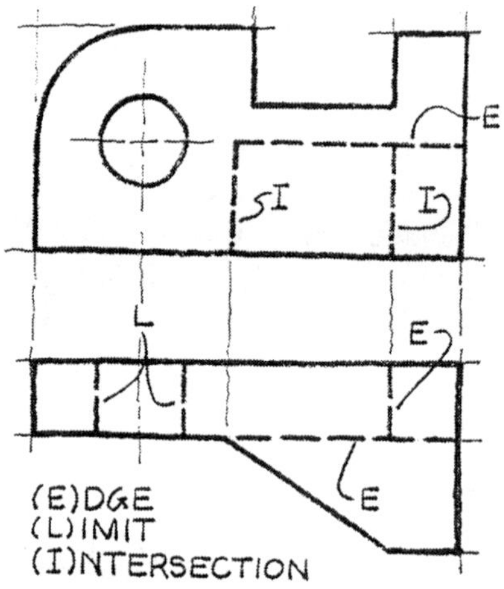

Fig. 4-4. Dashed lines on a sketch.

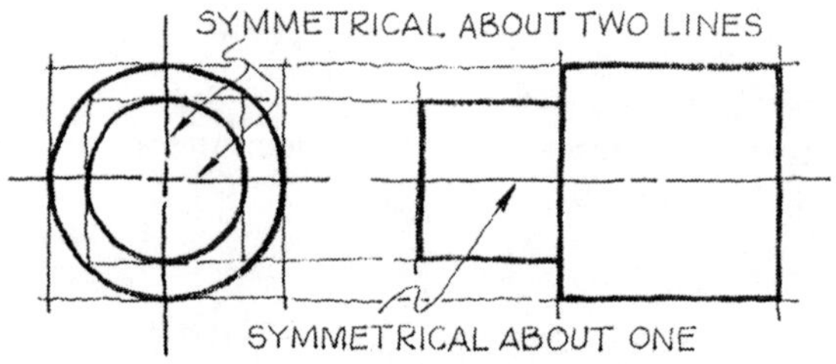

Fig. 4-5. Lines of symmetry.

2. Rough out the overall dimensions of the views, paying attention to symmetry. Keep the views aligned.

3. Add the detail and work *between the views*. That is, you may have to simultaneously work on two views, adding from one to another until they are complete.

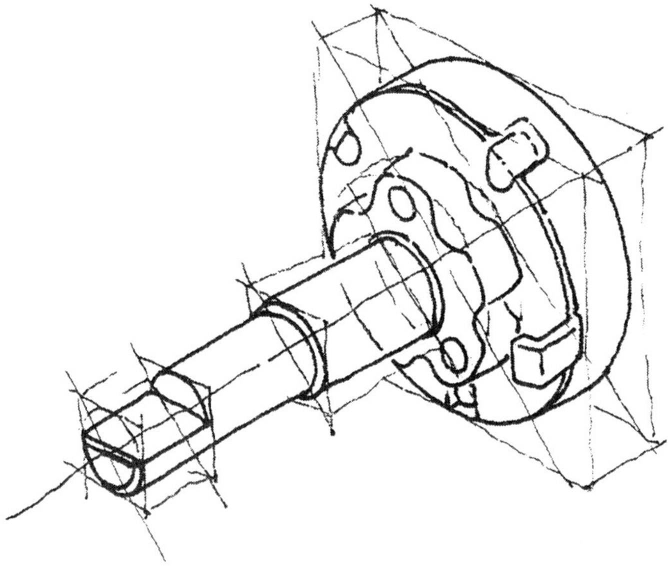

Fig. 4-6. Pictorial sketch.

CAD Example

A designer will work from pictorial to orthographic sketches in order to produce a CAD drawing. The overall strategy is much the same: Define the overall geometry first and then add the detail.

Figure 4-8 shows the same electrical part in its rough layout (a), edited layout (b), and detailed layout (c). Note, to simplify side view construction in (a), the UCS is relocated as necessary.

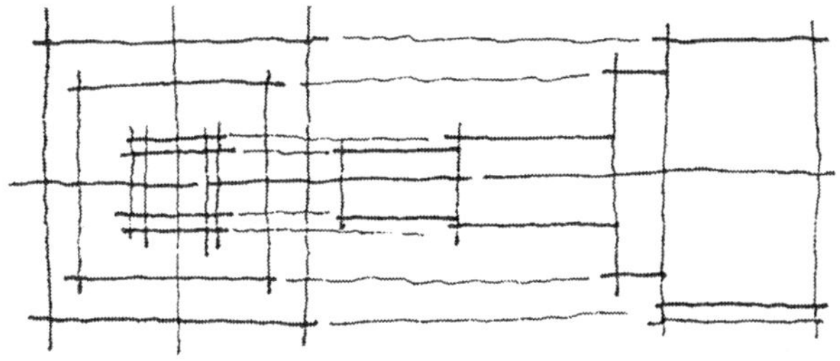

Step 1: Lay out overall dimensions.

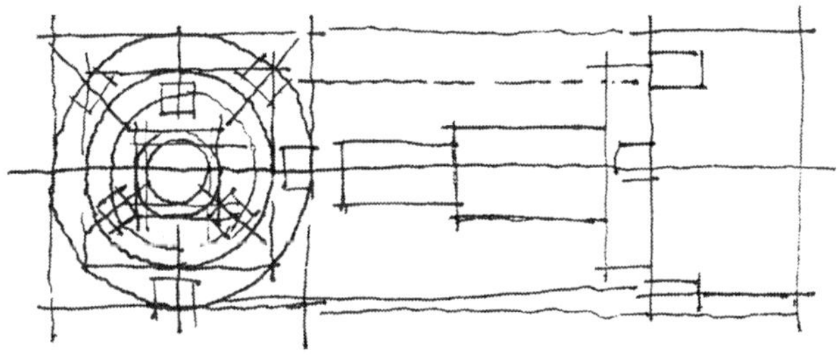

Step 2: Rough out details.

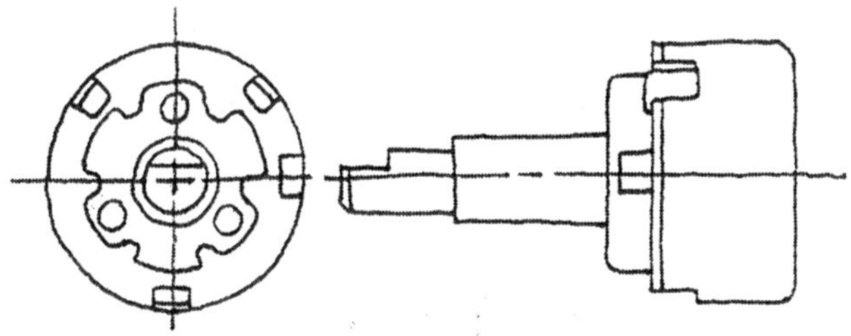

Step 3: Add detail.

Fig. 4-7. Steps in completing orthographic views.

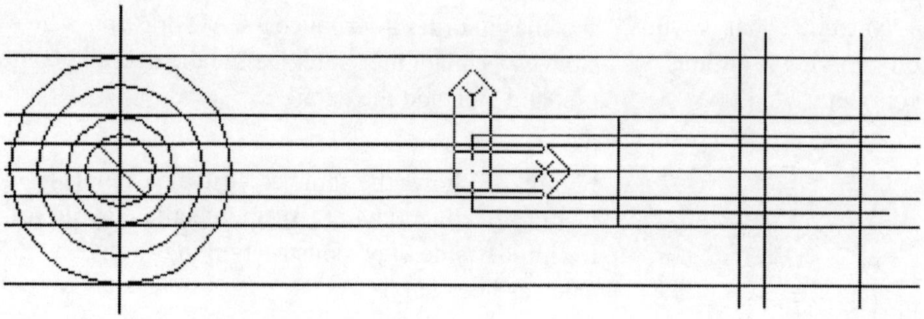

Fig. 4-8(a). Rough layout in CAD.

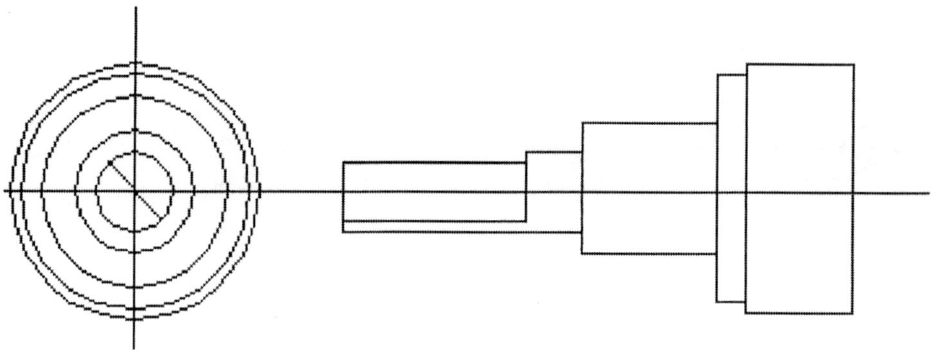

Fig. 4-8(b). Edited layout in CAD.

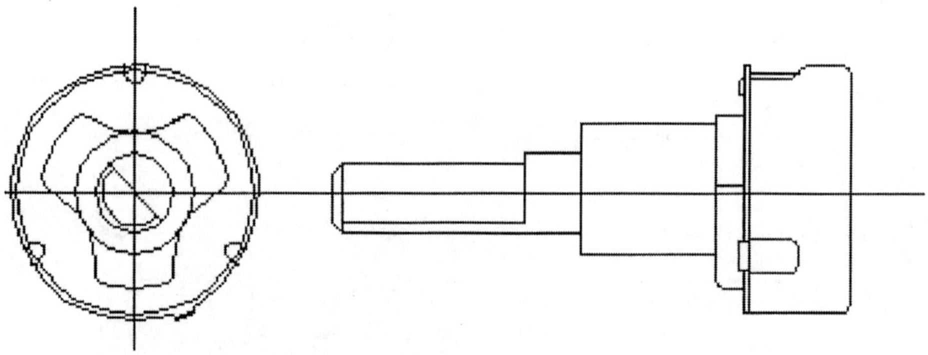

Fig. 4-8(c). Detailed layout in CAD.

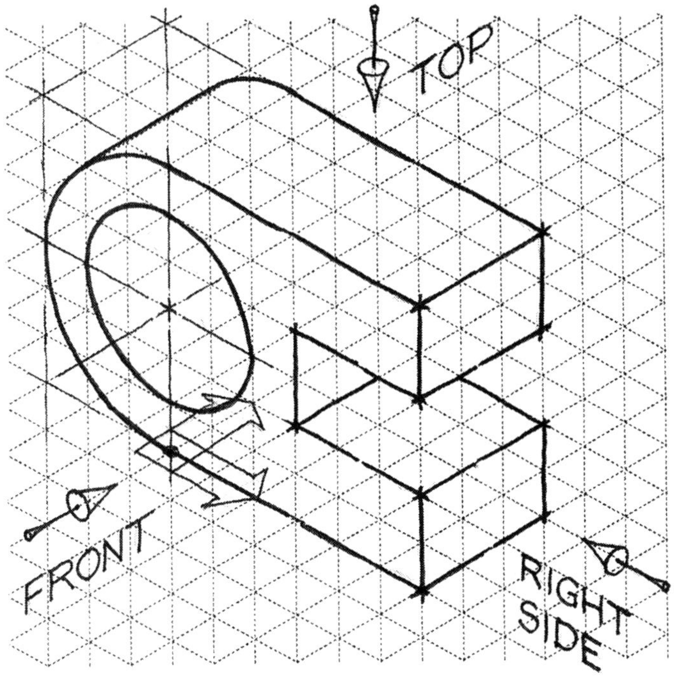

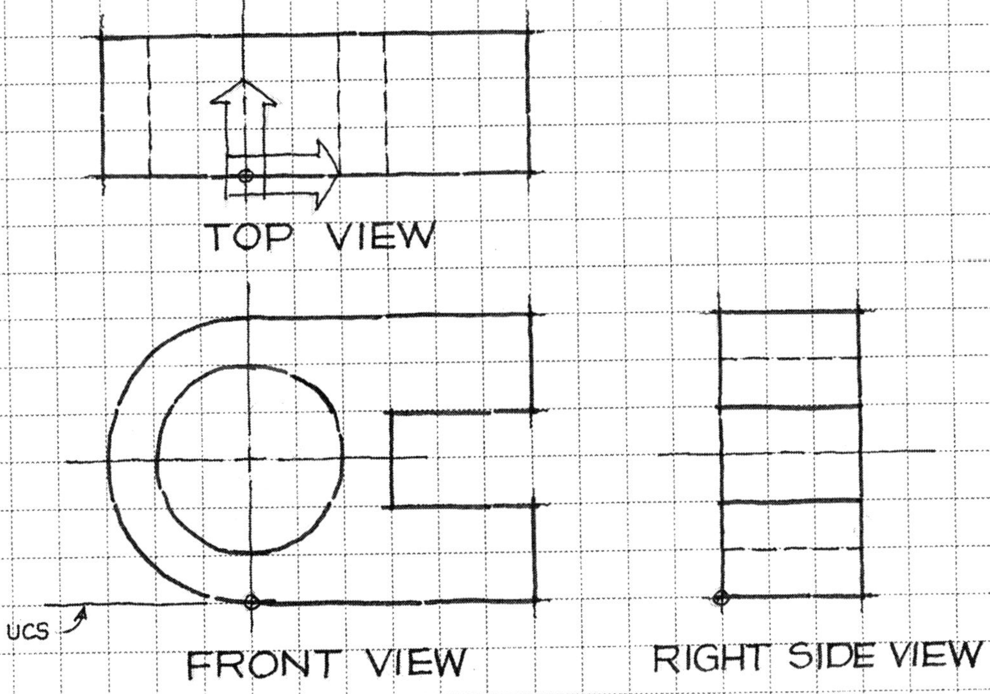

Drawn By:		FREEHAND SKETCHING	
	Scale	TITLE SAMPLE	
Date	Sheet of	Drawing Number A	4.0

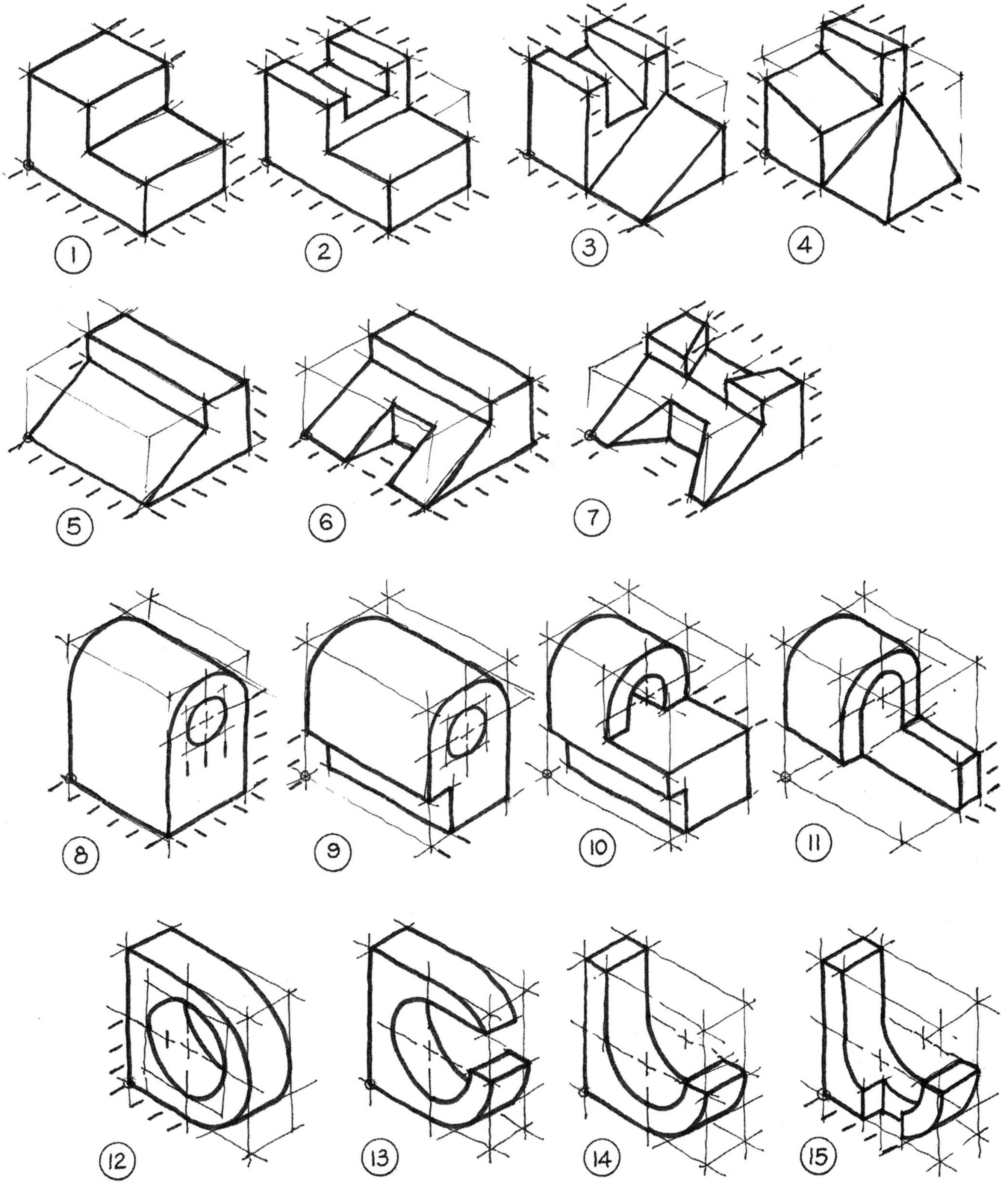

Drawn By:		FREEHAND SKETCHING	
	Scale	TITLE PICTORIAL OBJECTS	
Date	Sheet of	Drawing Number A	4.1

COMPLETE THREE-VIEW ORTHOGRAPHIC SKETCHES OF THE FOLLOWING OBJECTS. USE ORTHO GRID PAPER.

(A)

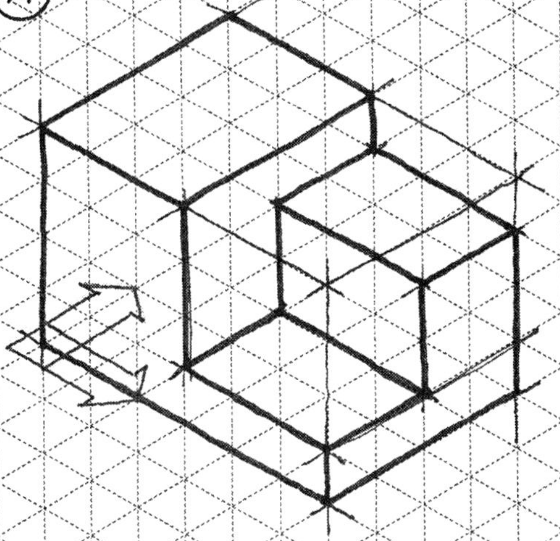

(B)

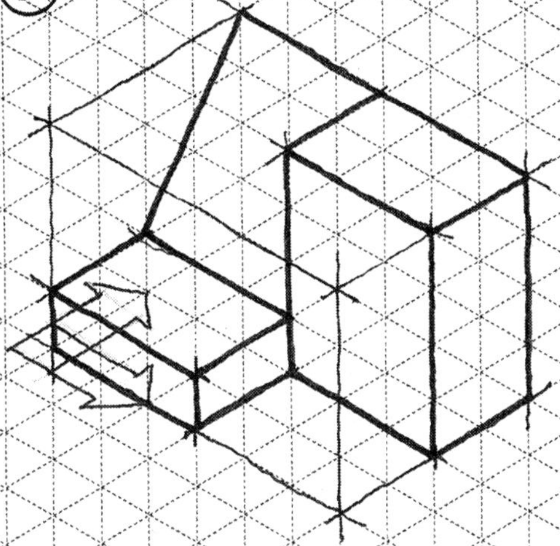

(C)

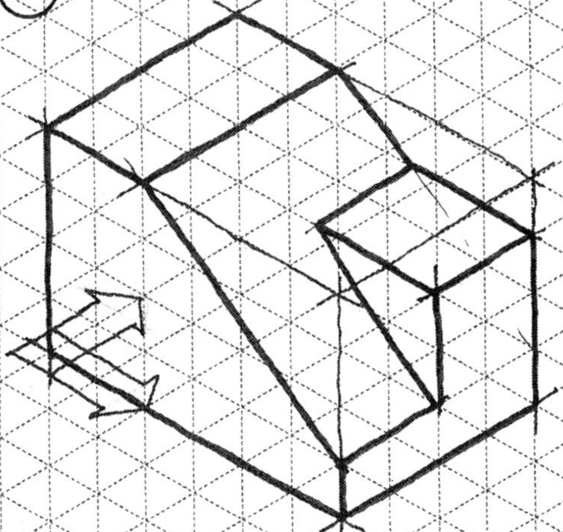

(D)

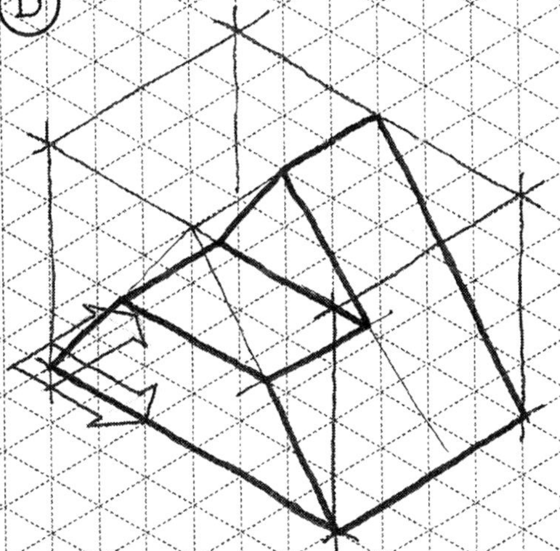

Drawn By:		FREEHAND SKETCHING	
	Scale	TITLE PICTORIAL OBJECTS	
Date	Sheet of	Drawing Number A	4.2

COMPLETE THREE-VIEW ORTHOGRAPHIC SKETCHES OF THE FOLLOWING OBJECTS. USE ORTHO GRID PAPER.

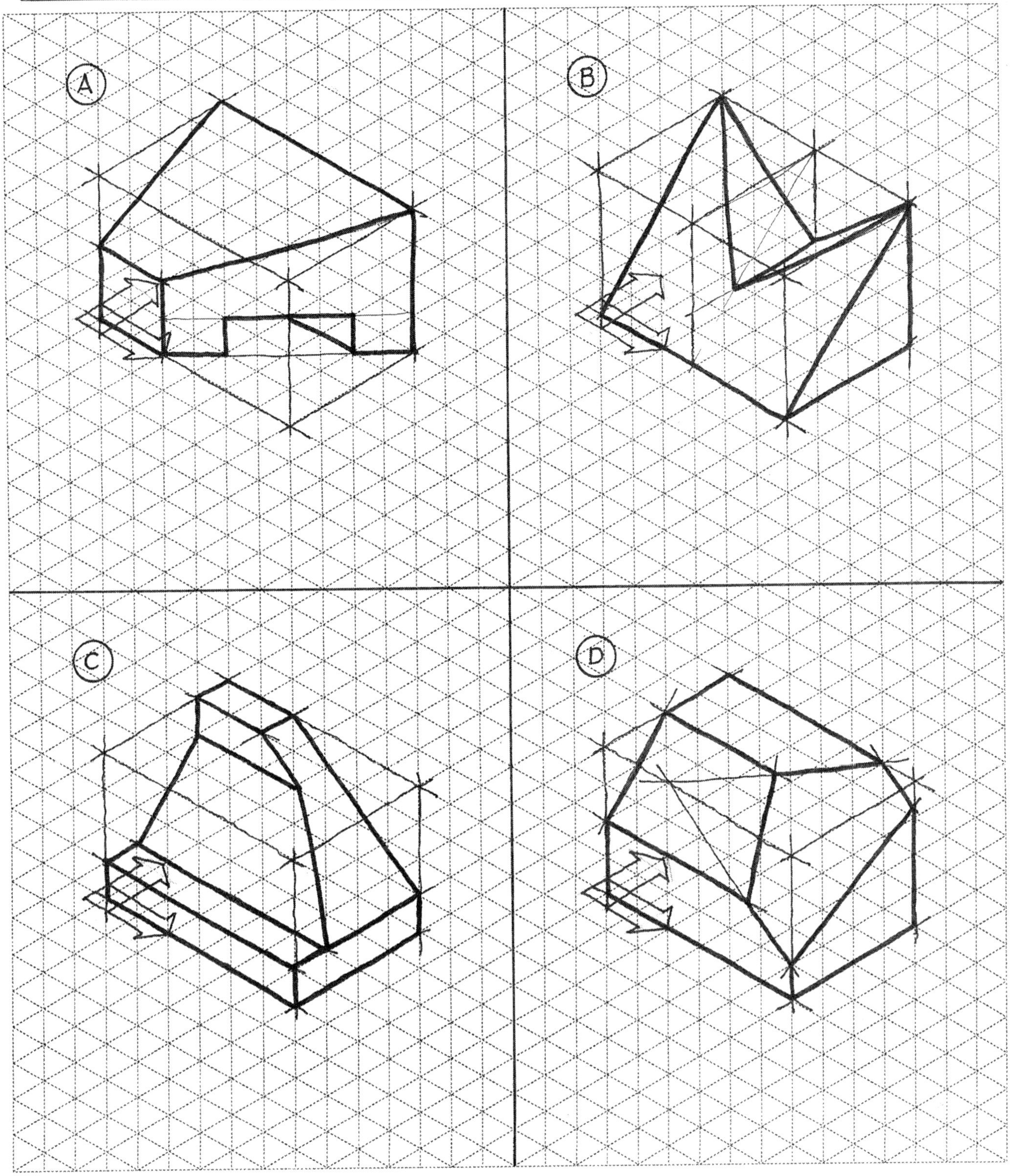

Drawn By:		FREEHAND SKETCHING	
	Scale	TITLE PICTORIAL OBJECTS	
Date	Sheet of	Drawing Number A	4.3

COMPLETE THREE-VIEW ORTHOGRAPHIC SKETCHES OF THE FOLLOWING OBJECTS. USE ORTHO GRID PAPER.

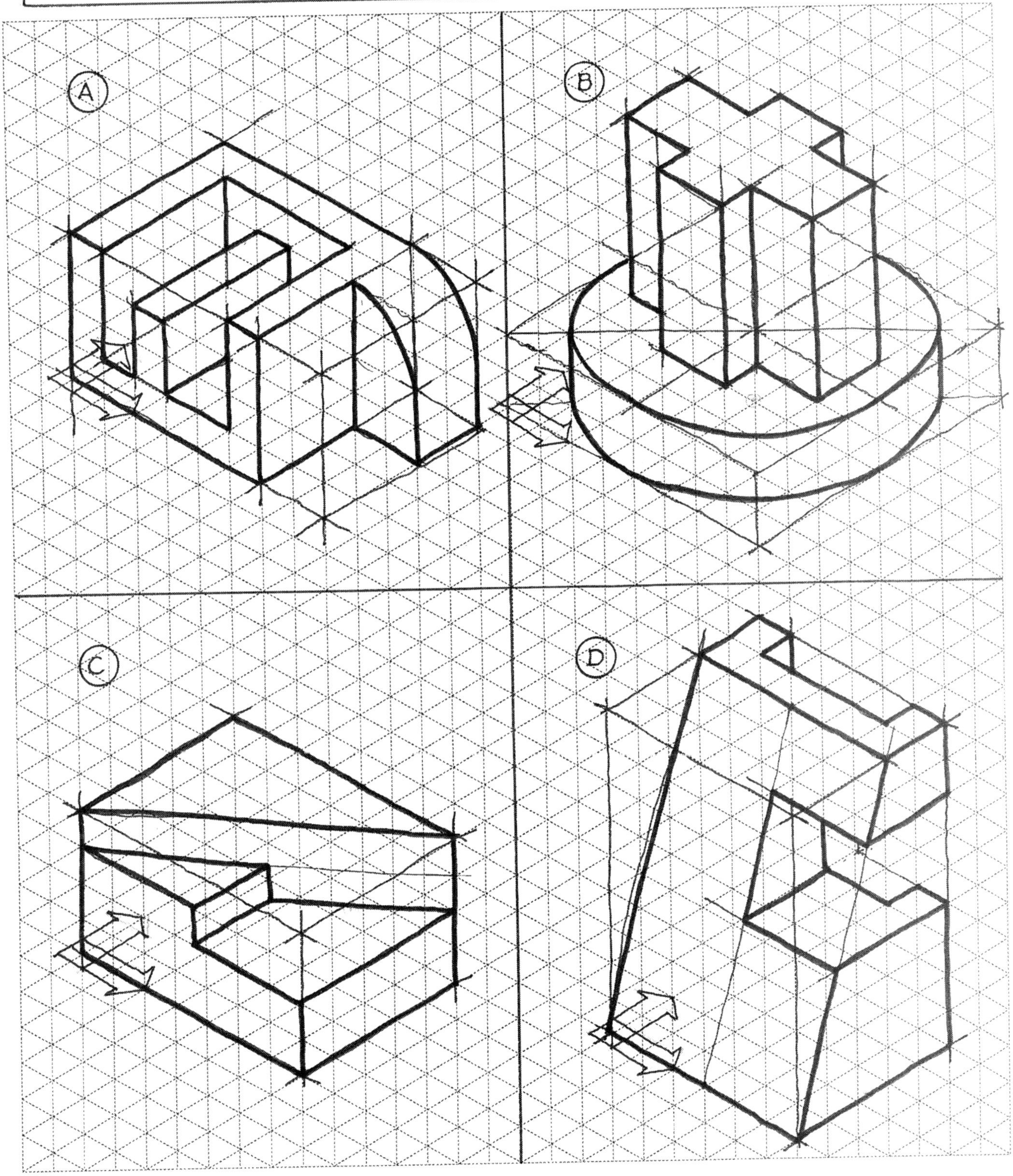

Drawn By:		FREEHAND SKETCHING	
	Scale	TITLE PICTORIAL OBJECTS	
Date	Sheet of	Drawing Number A	4.4

COMPLETE THREE-VIEW ORTHOGRAPHIC SKETCHES OF THE FOLLOWING OBJECTS. USE ORTHO GRID PAPER.

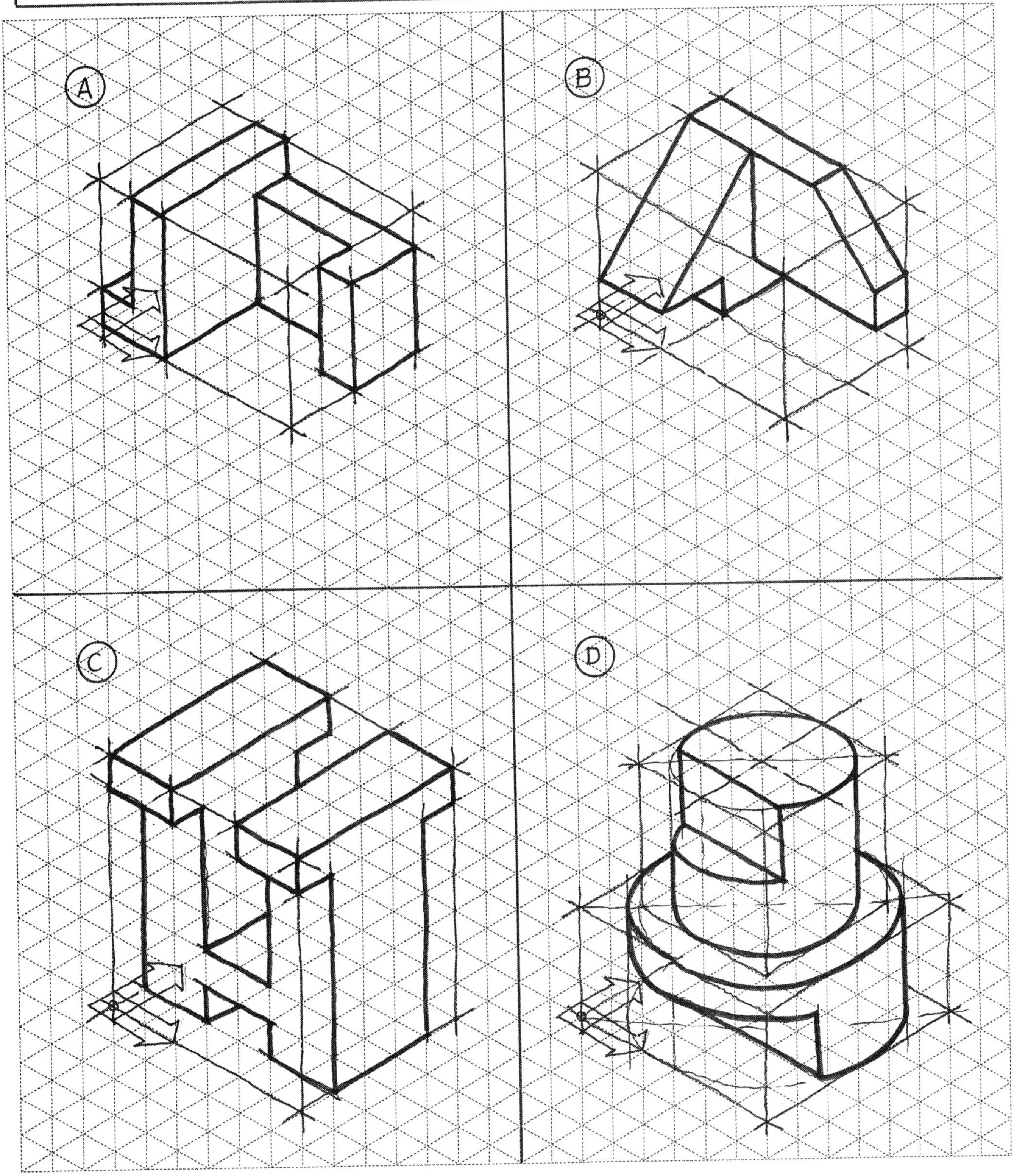

Drawn By:		FREEHAND SKETCHING	
	Scale	TITLE PICTORIAL OBJECTS	
Date	Sheet of	Drawing Number A	4.5

COMPLETE THREE-VIEW ORTHOGRAPHIC SKETCHES OF THE FOLLOWING OBJECTS. USE ORTHO GRID PAPER.

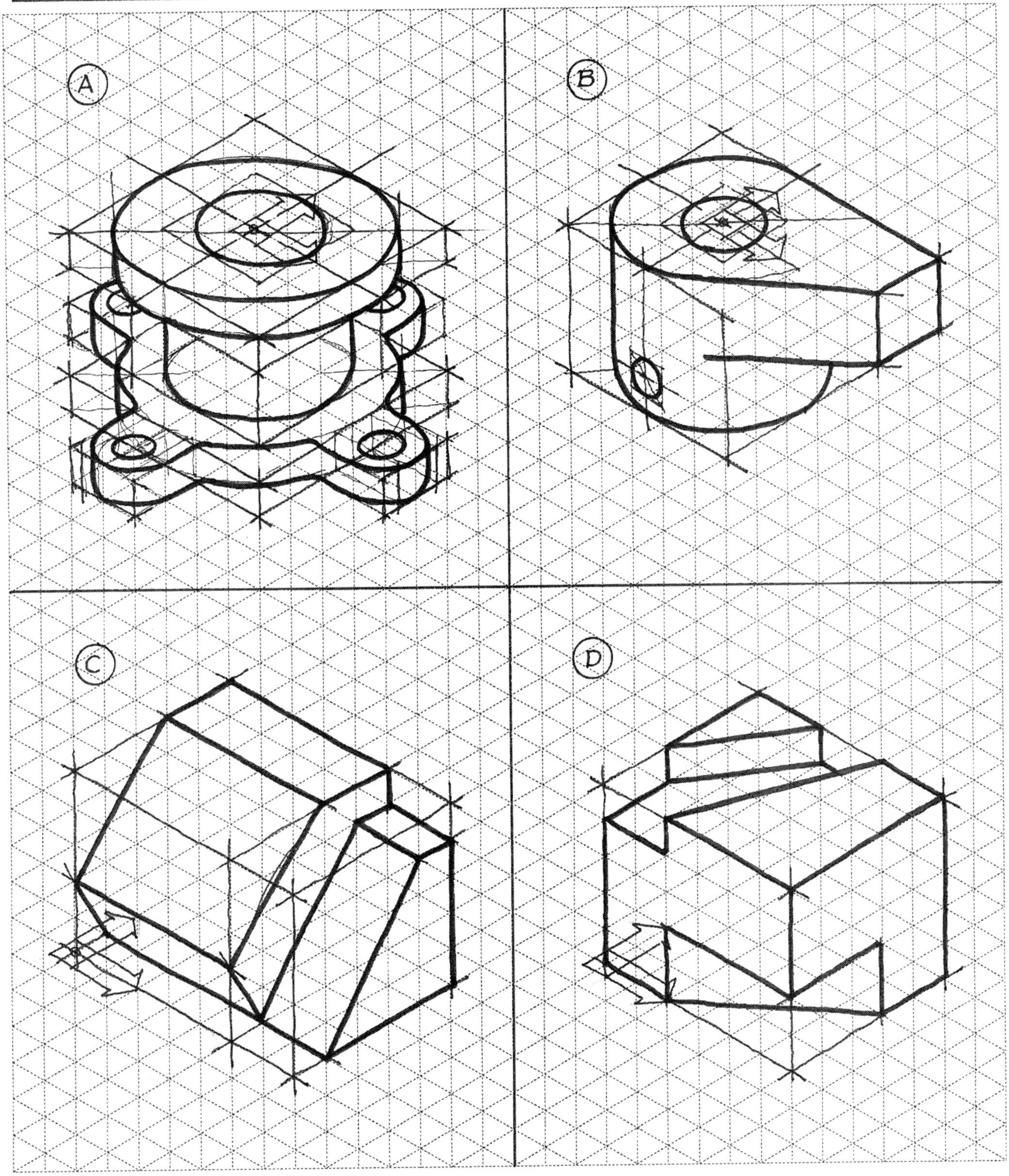

Drawn By:		FREEHAND SKETCHING	
	Scale	TITLE PICTORIAL OBJECTS	
Date	Sheet of	Drawing Number A	4.6

QUESTIONS:

1. What is the name of the object?
2. What line represents surface (G) in the top view?
3. What is dimension (J)?
4. What is the total width of the part as shown by dimension (A)?
5. What is the radius of rounded end (R)?
6. From what type of material is the part made?
7. What is the size of dimension (C)?
8. Which line in the top view is used to represent surface (K) in the right side view?
9. What type of line is (D)?
10. What type of line is (M)?

ANSWERS:

1. ____________
2. ____________
3. ____________
4. ____________
5. ____________
6. ____________
7. ____________
8. ____________
9. ____________
10. ____________

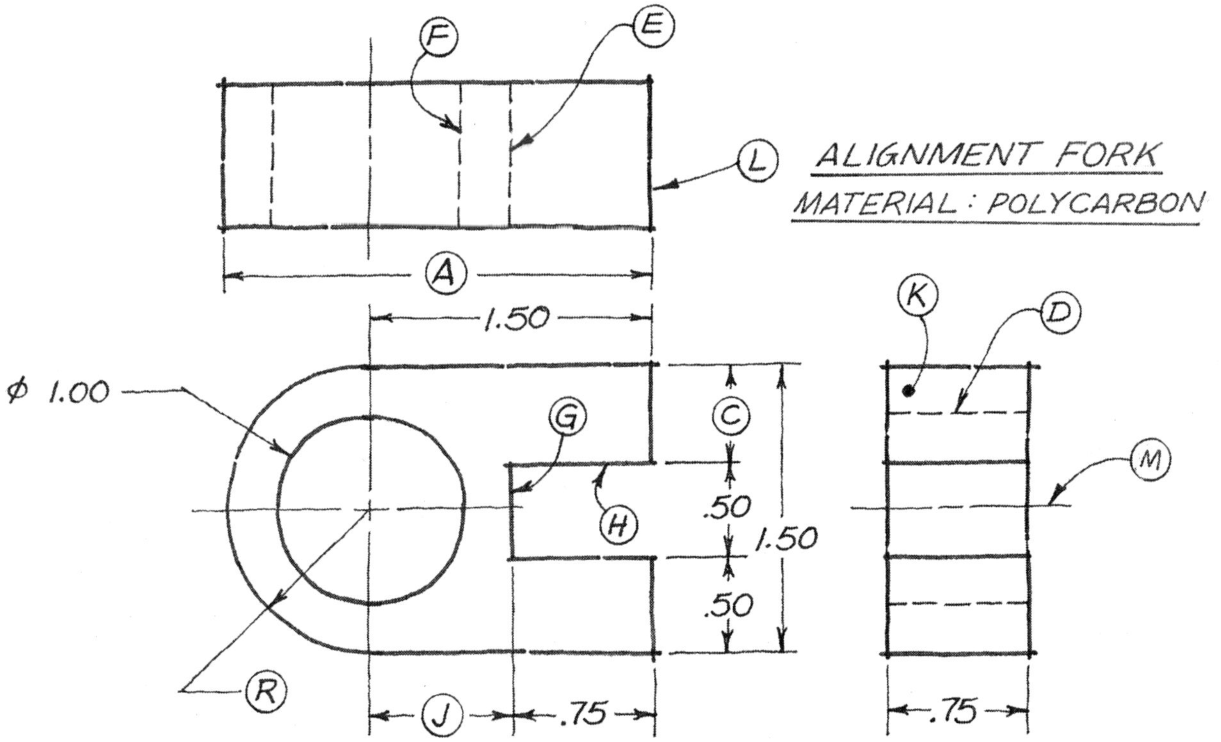

Drawn By:		FREEHAND SKETCHING	
	Scale	TITLE READING DRAWINGS	
Date	Sheet of	Drawing Number A	4.7

5 Isometric Sketching

Coordinated with VMGED Chapter 12

Isometric Views—General Conditions

An *isometric view* (iso meaning equal, metric meaning measure) displays three sides of an object in a *pictorial* orthographic view. The three sides may be top, front, and one side, or any other combination of three mutually perpendicular faces (Fig. 5-1). Note that one view presents the isometric in a slightly different orientation.

An isometric view displays the three faces in equal inclination and scale. Measurements can be made along or parallel to any of the isometric axes. These axes are set at 30, 90, and 150 degrees on the drawing sheet. Measurements cannot be directly made at any other angle (Fig. 5-2).

Isometric sketches usually don't include hidden features by means of hidden or dashed lines.

A true isometric projection (as in 3D CAD) equally foreshortens measurements along the axes to roughly 82 percent. To make construction simple, the usual practice in isometric drawing is to measure along isometric axes at full scale. This produces a correctly proportioned, but slightly larger view than a projection.

Sheet 5.0 on page 85 shows the steps in making an isometric sketch. The overall height, width, and depth are noted as distances D1, D2, and D3.

Step 1. Place the most descriptive orientation to the front so an appropriate side is shown. Construct the overall shape of the object using D1, D2, and D3. This is called the *boxing-out* construction method.

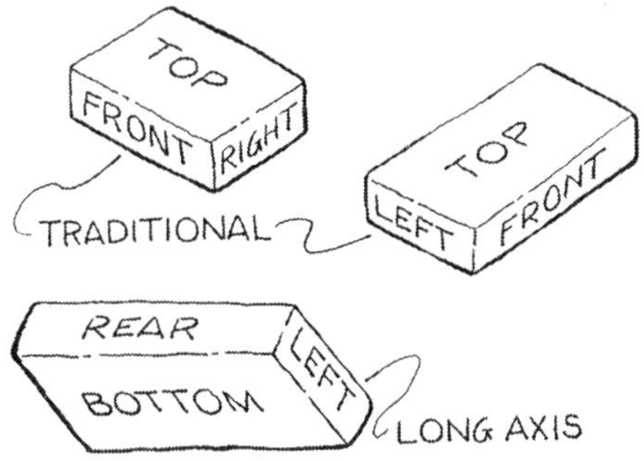

Fig. 5-1. Arrangement of isometric views.

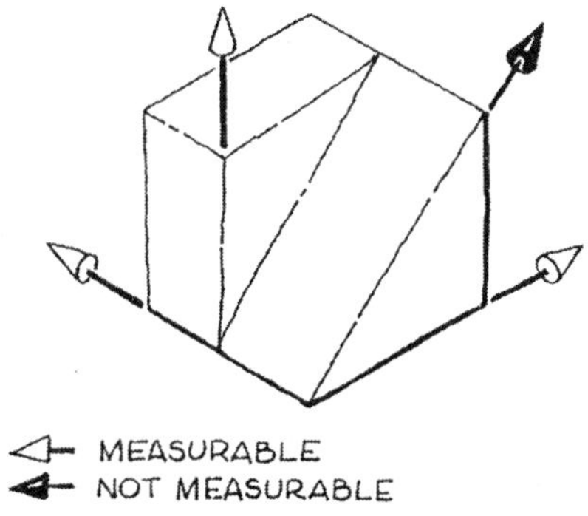

Fig. 5-2. Valid measurement directions in isometric.

Step 2. Locate the edges and surfaces by measuring parallel to the isometric axes. Do this feature by feature.

Step 3. Complete the pictorial by darkening the visible edges and surfaces.

Isometric and Non-Isometric Lines

An *isometric line* is any line parallel to one of the isometric axes. A *non-isometric line* is any line *not* parallel to an isometric axis. Isometric lines can be measured directly. Non-isometric lines must be found by measuring components of the right triangle of which the non-isometric line is the hypotenuse (Fig. 5-3). This is called the *offset method* of construction.

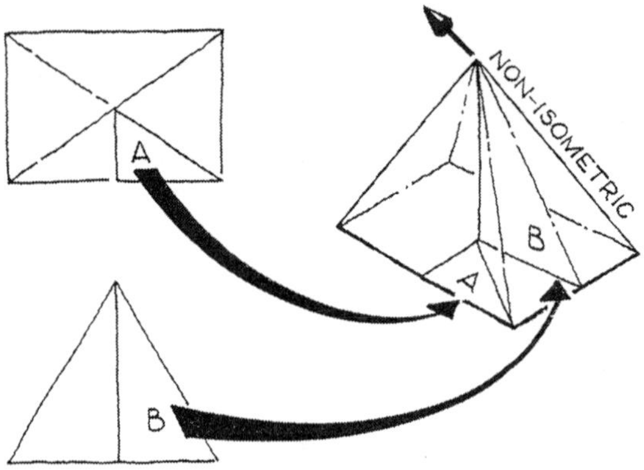

Fig. 5-3. Offset method of constructing non-isometric lines.

Inclined and Oblique Planes

An *inclined plane* (Fig. 5-4) may be considered an isometric plane revolved about one of the isometric axes. Two boundary lines are non-isometric lines and must be found as such. Note that two boundary lines are isometric lines and can be measured directly.

An *oblique plane* has been rotated about two isometric axes (Fig. 5-5). This results in all boundary lines being non-isometric. You cannot measure directly on an oblique plane.

Assignment: On the completed isometric (Step 3 of Sheet 5.0), color all frontal planes blue; color all profile planes green; color all horizontal planes yellow; color all inclined planes gray; color all oblique planes red.

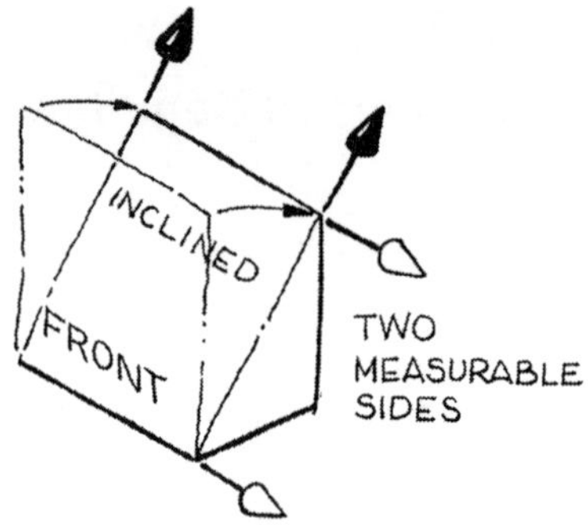

Fig. 5-4. Inclined plane in isometric.

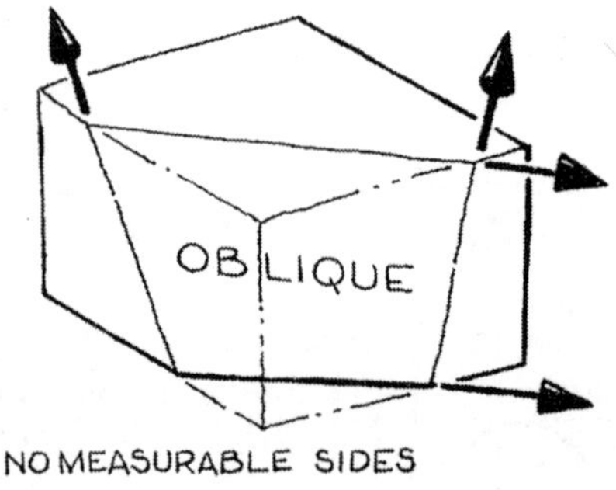

Fig. 5-5. Oblique plane in isometric.

Angles in Isometric

Because an angle other than horizontal or vertical creates a non-isometric line, it is found by the offset method. Sometimes, sketching an angle first orthographically and then transferring the right angle legs into isometric is the easiest construction method (Fig. 5-6).

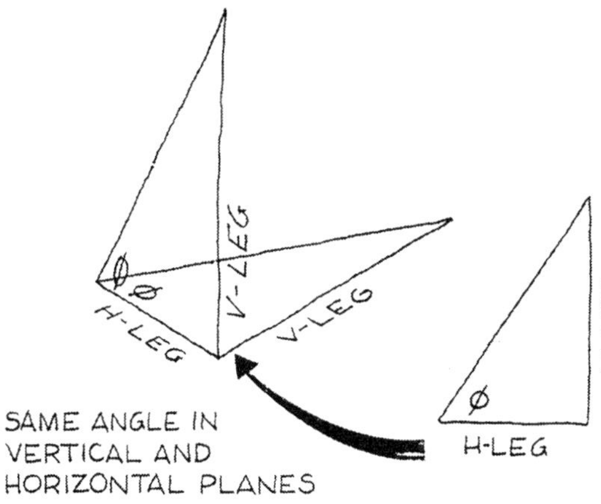

Fig. 5-6. An angle in isometric.

Circular Features

The general method for sketching any curve in isometric is a modified offset construction—*grid construction*. Both circular and non-circular curves can be done this way (Fig. 5-7). Irregular curves require a grid with divisions fine enough to produce a sufficient number of control points so the curve can be plotted. Sketching circles involves the use of a simplified grid method where the shape is boxed out to find the axes (diameters) and tangent points (Fig. 5-8).

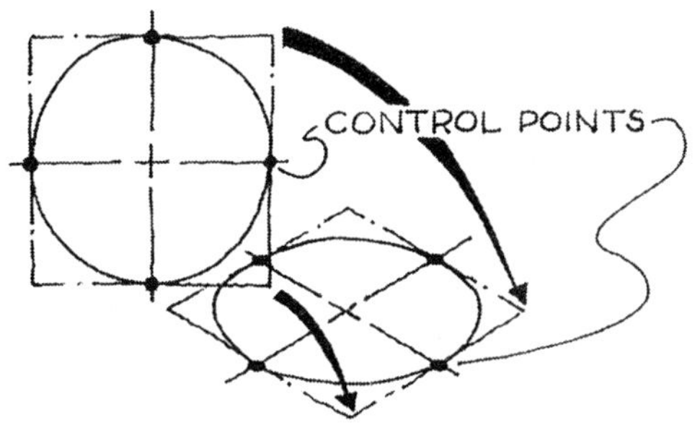

Fig. 5-8. Circular construction by boxing out.

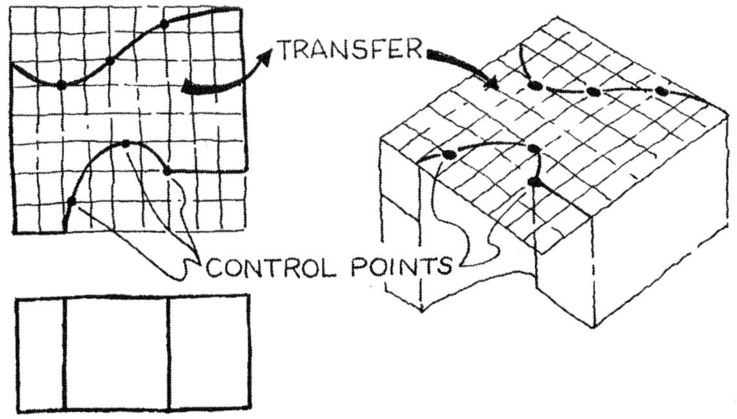

Fig. 5-7. Grid method of constructing curves.

Practical Example

In Fig. 5-9, two orthographic views of an object having circular and inclined features are shown. A solution in isometric requires the following steps:

Step 1. Decide on the orientation for a front isometric face. This can be in any of the isometric faces but care should be taken not to obscure any important features.

Step 2. Box out the overall dimensions of height, width, and depth.

Step 3. Locate the inclined surface by the offset method.

Step 4. Locate the center of circular features by measuring parallel to isometric axes. Draw the center lines and box out the unitary form (rhombus) that encloses each ellipse.

Step 5. Complete linework to show visibility.

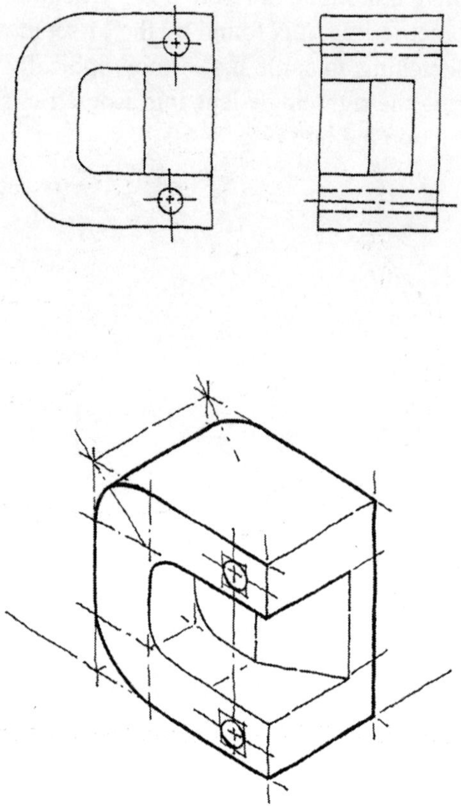

Fig. 5-9. Practical example of an isometric sketch.

CAD Example—2D

The same sketching techniques of boxing out, offset, and grid construction can be applied using 2D CAD tools (Fig. 5-10). The isometric drawing is made on the X-Y construction plane. Some of these tools automate construction at isometric angles (30, 90, and 150 degrees) and the exposure and angle of isometric ellipses.

CAD Example—3D

An *isometric projection* can be achieved in the 3D CAD front view by rotating the object—45 degrees about the vertical or Y-axis and 35.266 degrees about the horizontal or X-axis. Or, a new view can be created by moving the viewer (camera) an equal distance (say, 100 units) along the X-, Y-, and Z-axes and looking back at the origin.

In Fig. 5-11, note that the front view axis system has remained in place (the UCS centered on one of the circular features), signifying the viewer has taken a position at X=1, Y=1, Z=1, looking back at 0, 0, 0.

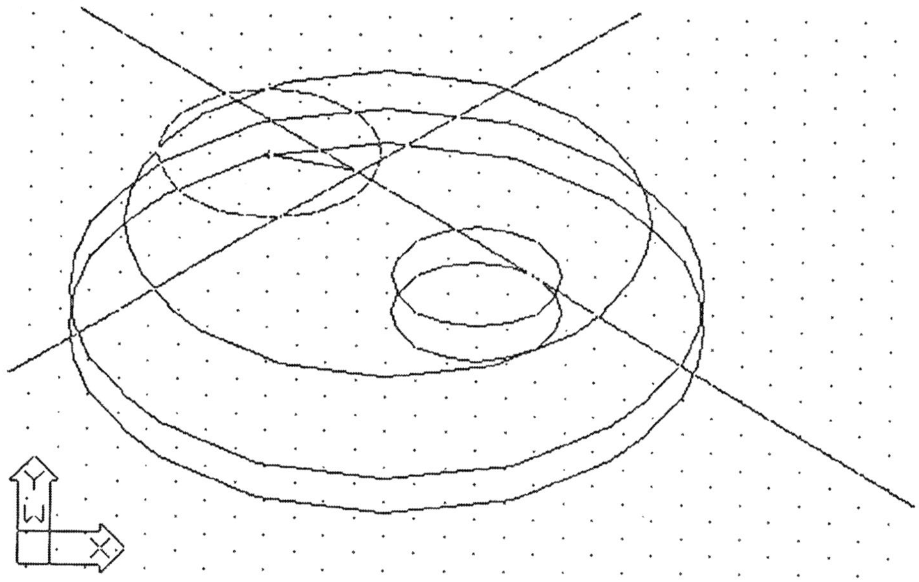

Fig. 5-10. Automated 2D CAD isometric constructions.

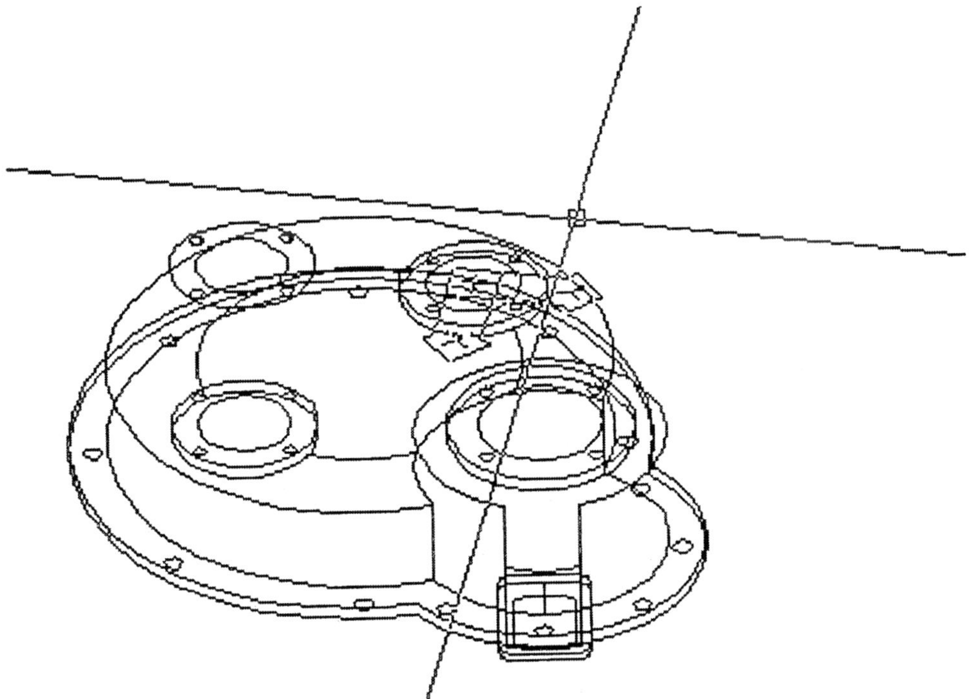

Fig. 5-11. Isometric by 3D viewpoint.

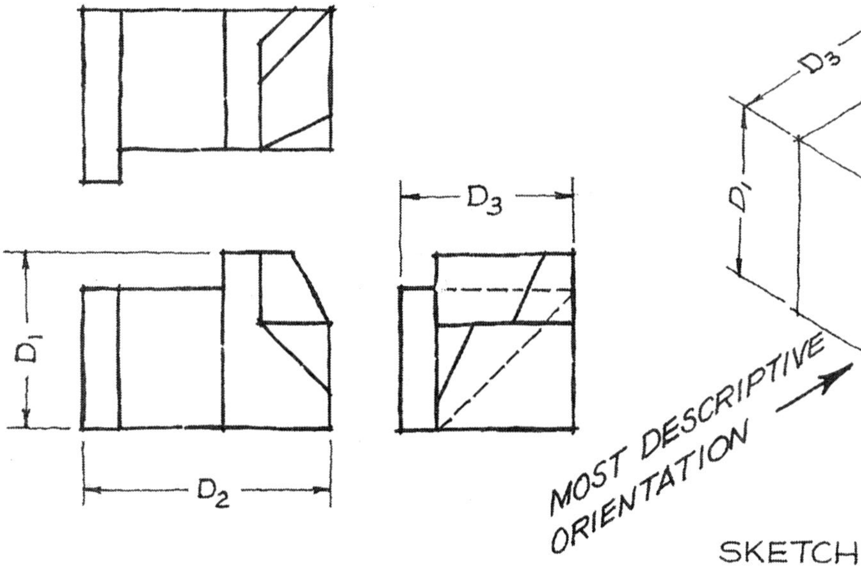

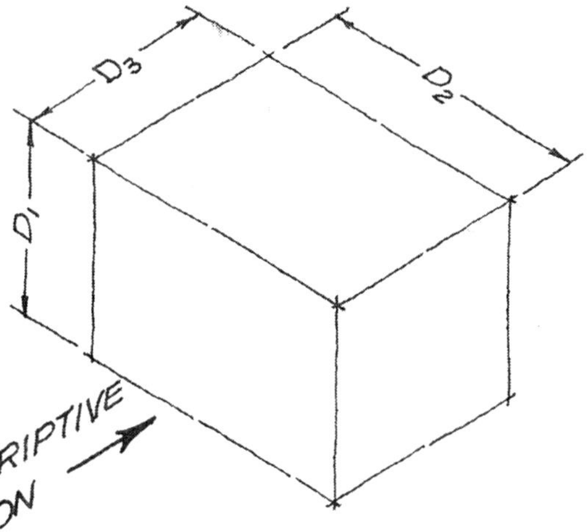

STEP 1

SKETCH ENCLOSING BOX

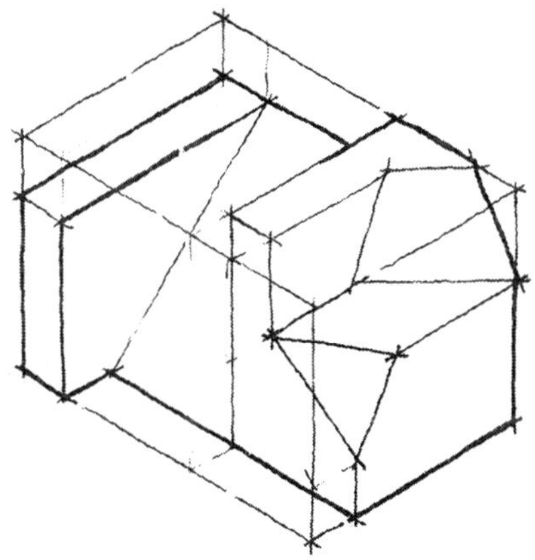

STEP 2

LOCATE EDGES AND SURFACES

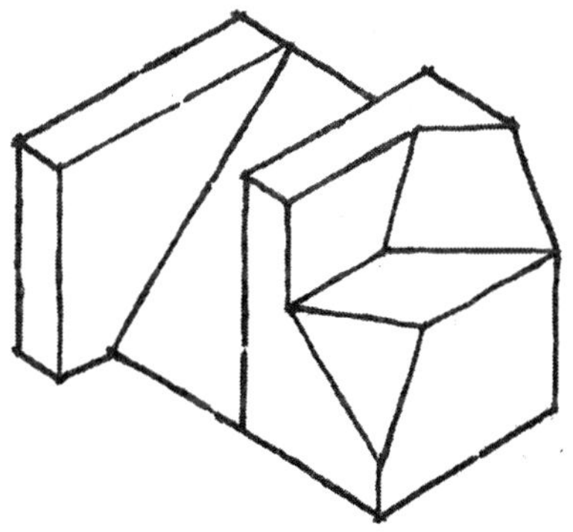

STEP 3

COMPLETE PICTORIAL

STEPS IN MAKING AN ISOMETRIC SKETCH

Drawn By:		FREEHAND SKETCHING	
	Scale	TITLE SAMPLE	
Date	Sheet of	Drawing Number A	5.0

COMPLETE AN ISOMETRIC SKETCH OF EACH OBJECT. SHADE INCLINED AND OBLIQUE PLANES AS ASSIGNED.

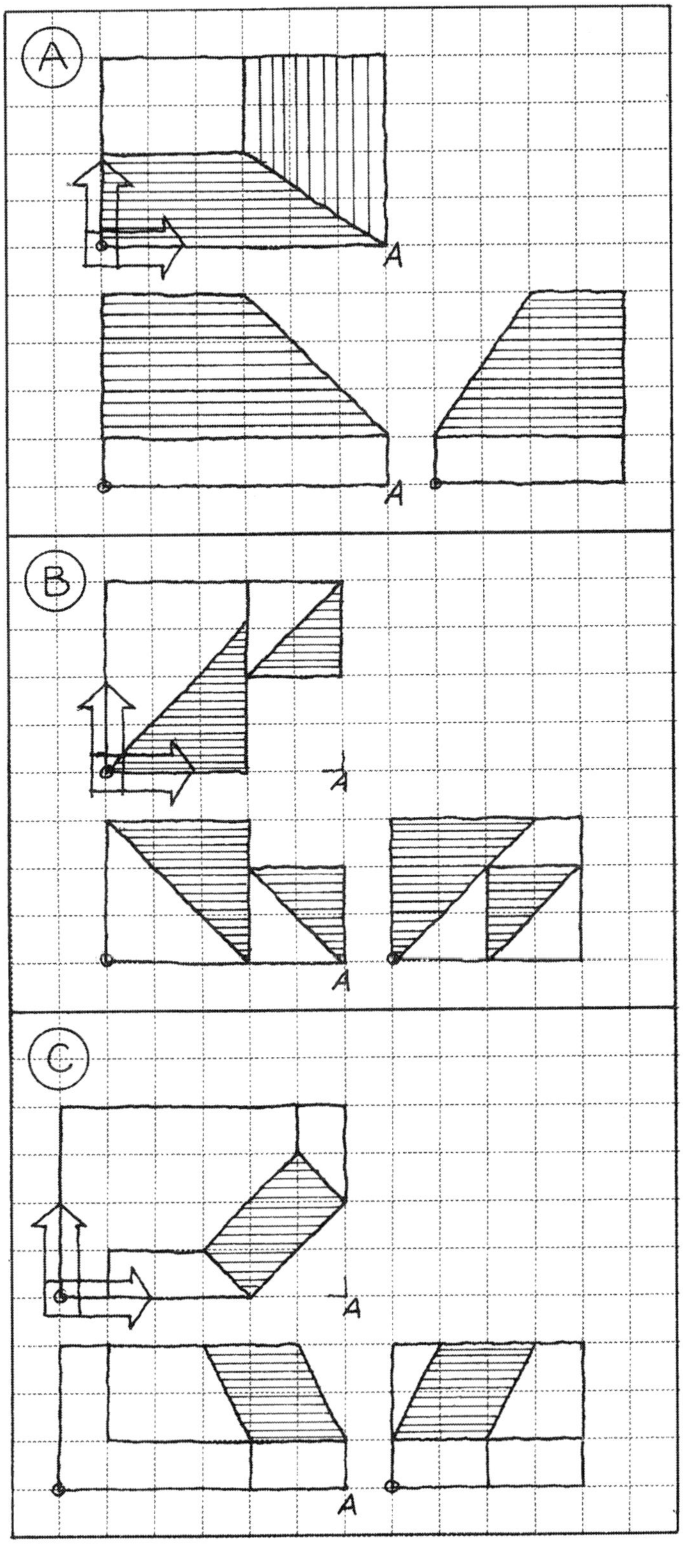

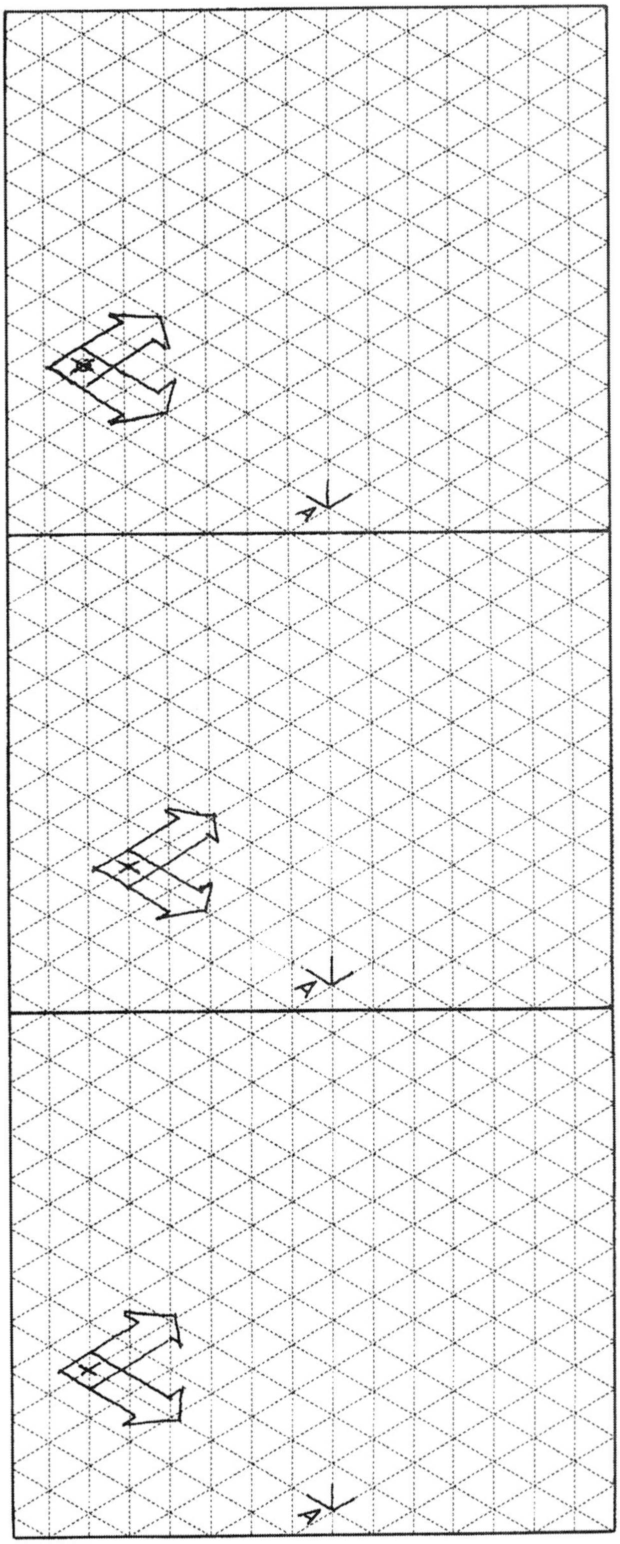

Drawn By:		FREEHAND SKETCHING	
	Scale	TITLE ISOMETRIC–PLANES	
Date	Sheet of	Drawing Number A	5.01

COMPLETE AN ISOMETRIC SKETCH OF THE FOLLOWING OBJECTS. CHOOSE THE MOST DESCRIPTIVE FEATURE. USE ISO GRID PAPER.

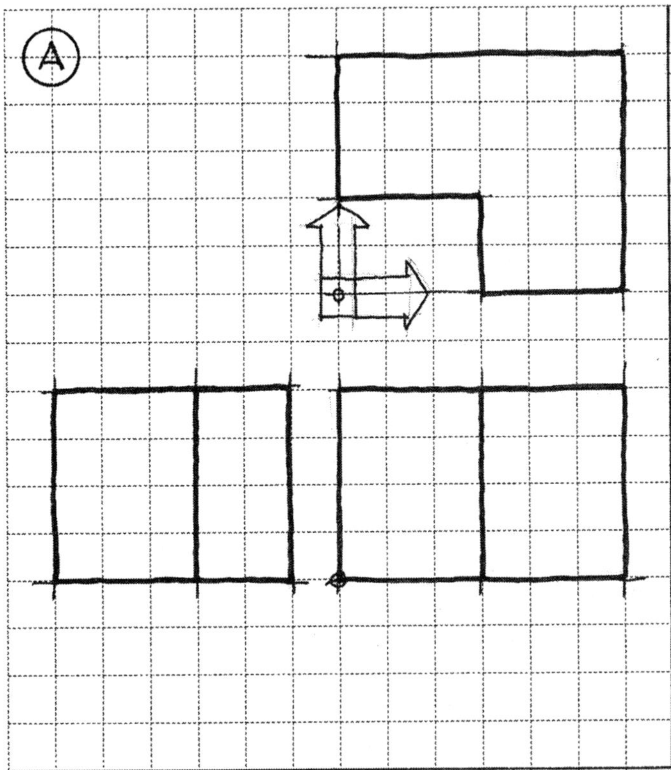

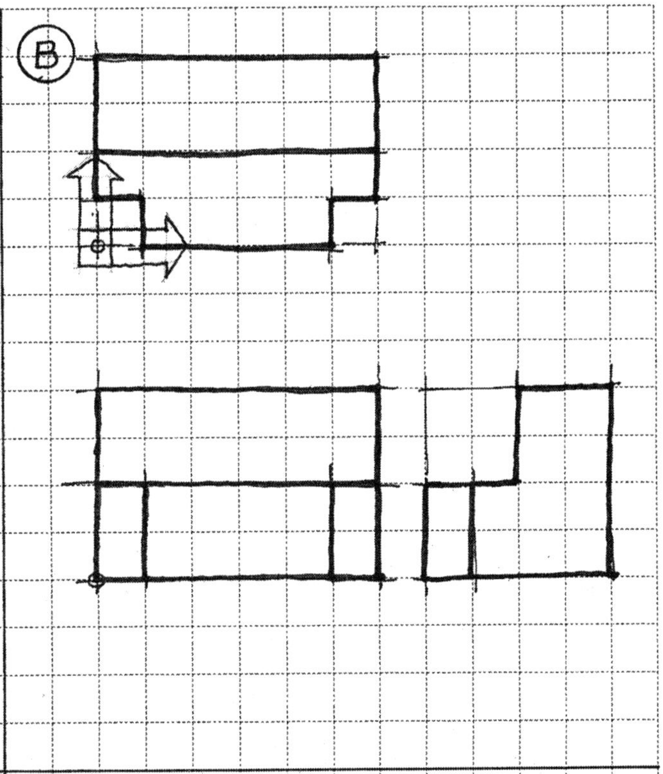

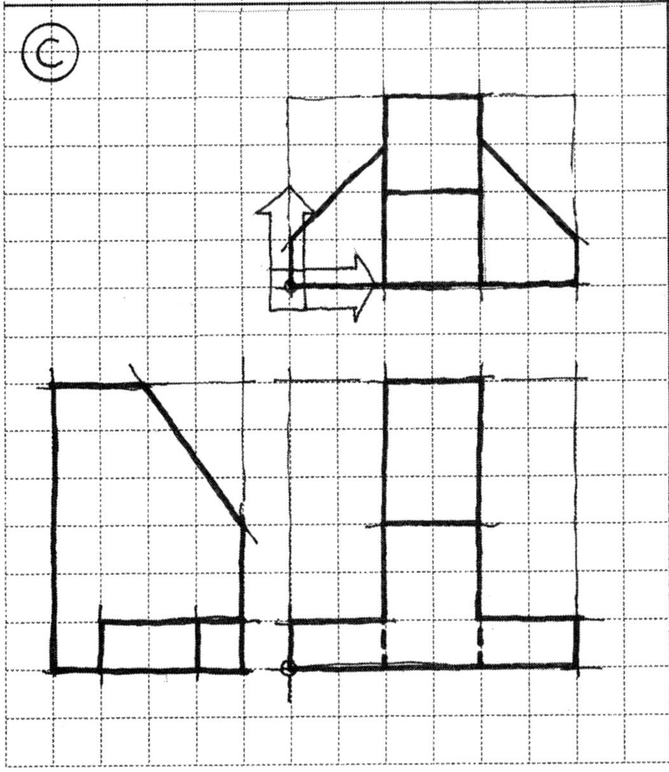

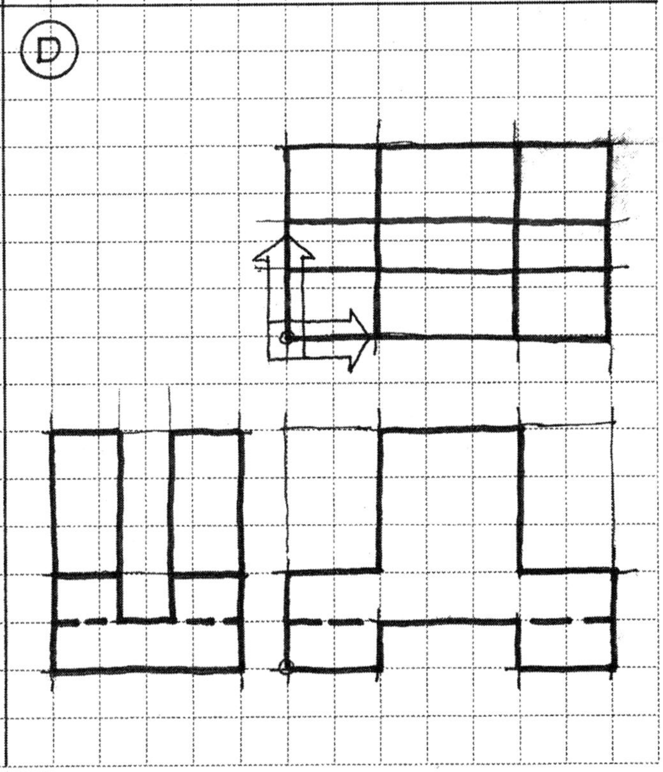

Drawn By:		FREEHAND SKETCHING	
	Scale	TITLE ORTHOGRAPHICS	
Date	Sheet of	Drawing Number A	5.1

COMPLETE AN ISOMETRIC SKETCH OF THE FOLLOWING OBJECTS.
CHOOSE THE MOST DESCRIPTIVE ORIENTATION. USE ISO GRID PAPER.

(A)

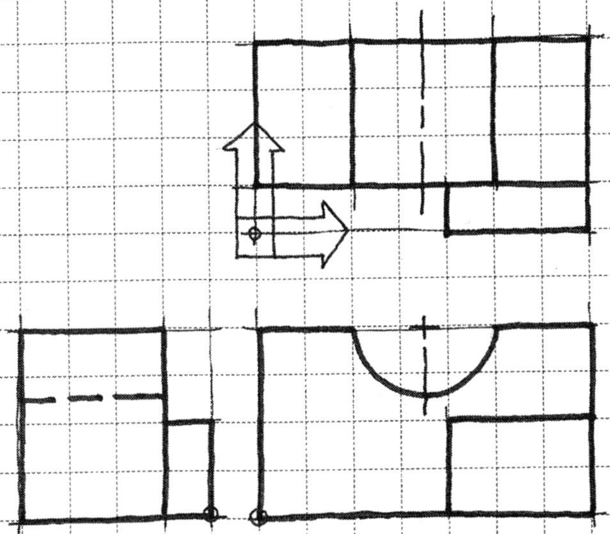

(B)

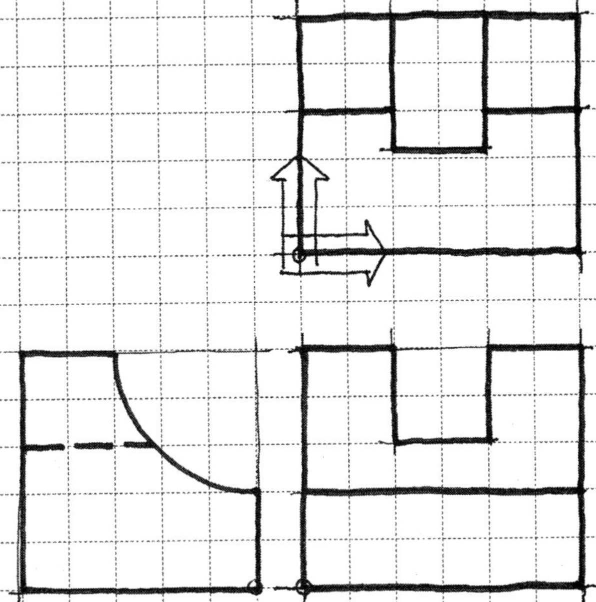

(C)

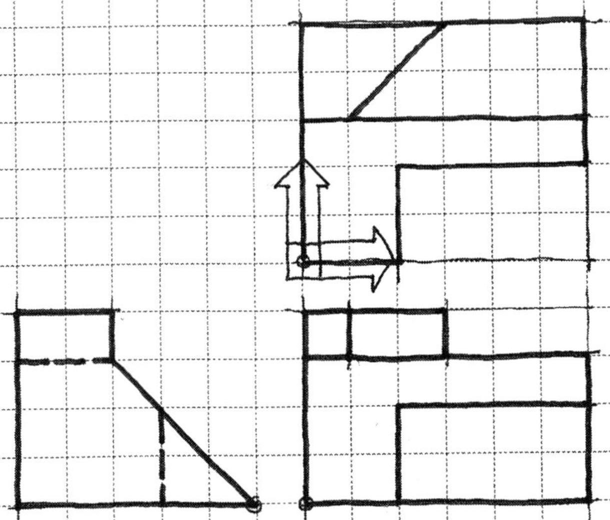

(D)

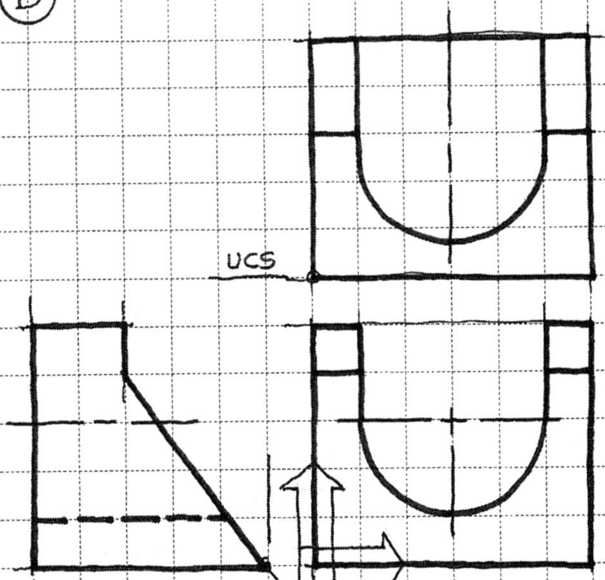

Drawn By:		FREEHAND SKETCHING	
	Scale	TITLE ORTHOGRAPHICS	
Date	Sheet of	Drawing Number A	5.2

COMPLETE AN ISOMETRIC SKETCH OF THE FOLLOWING OBJECTS. CHOOSE THE MOST DESCRIPTIVE ORIENTATION. USE ISO GRID PAPER.

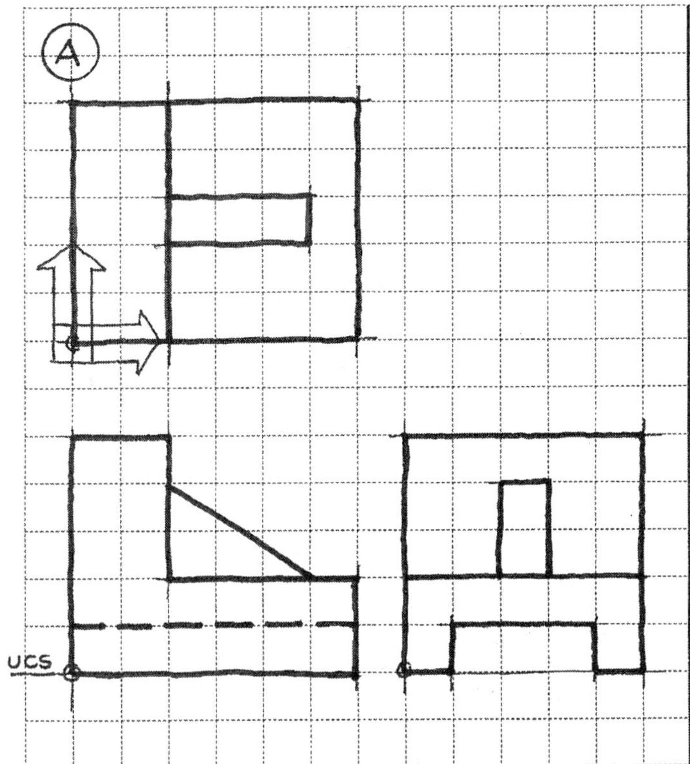

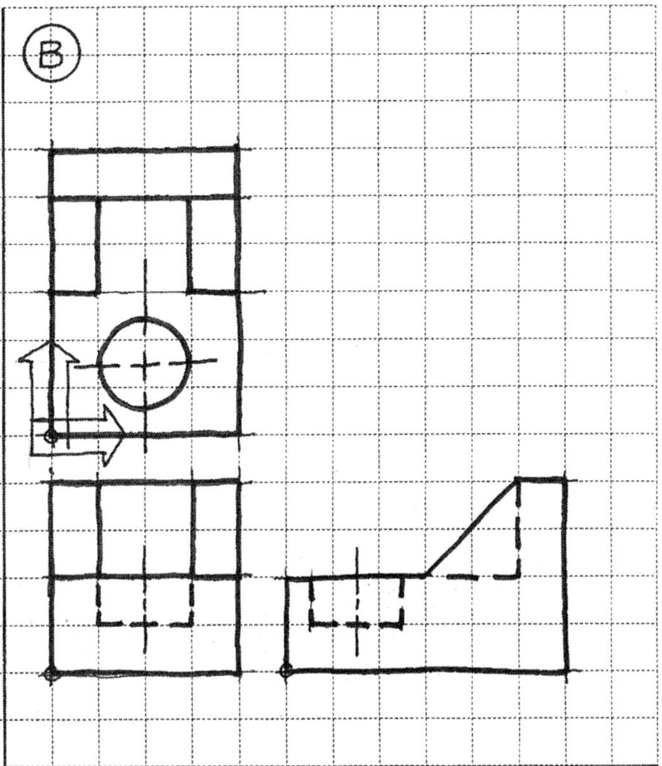

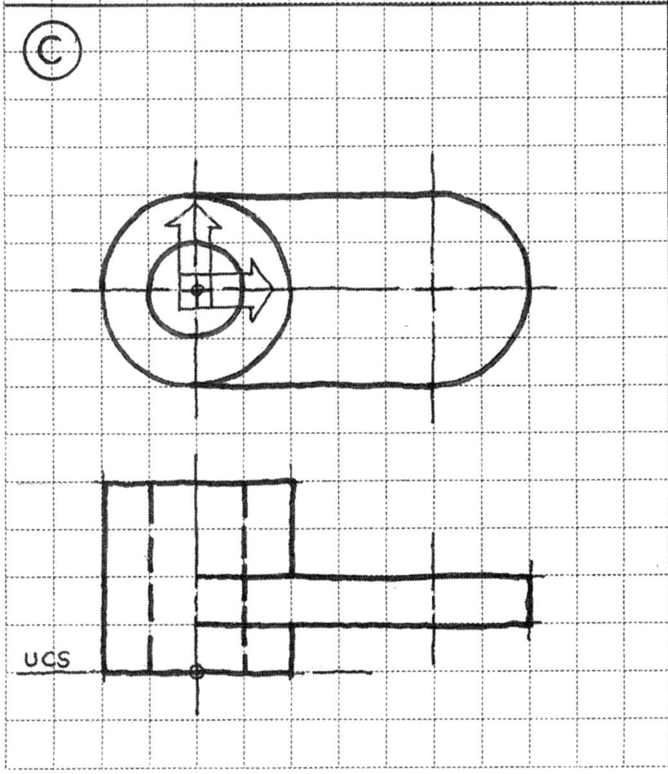

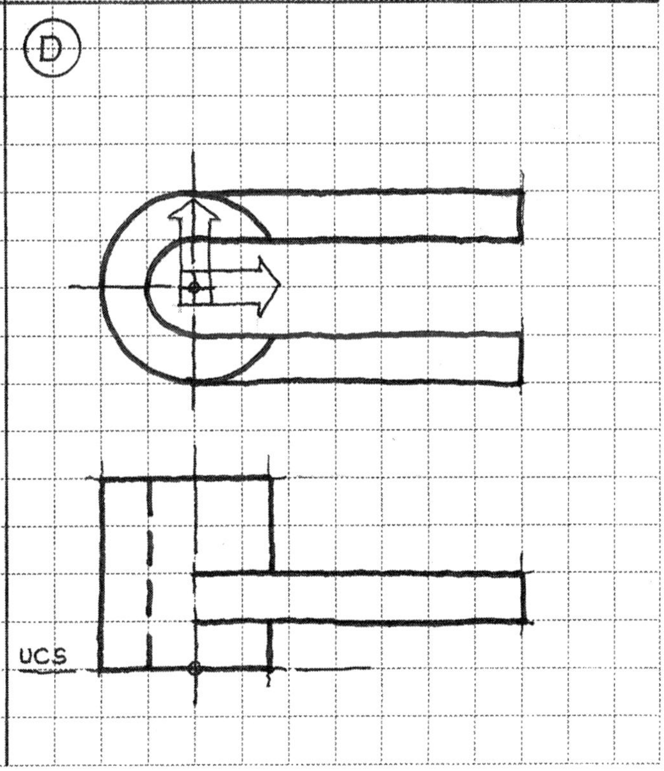

Drawn By:		FREEHAND SKETCHING	
	Scale	TITLE ORTHOGRAPHICS	
Date	Sheet of	Drawing Number A	5.3

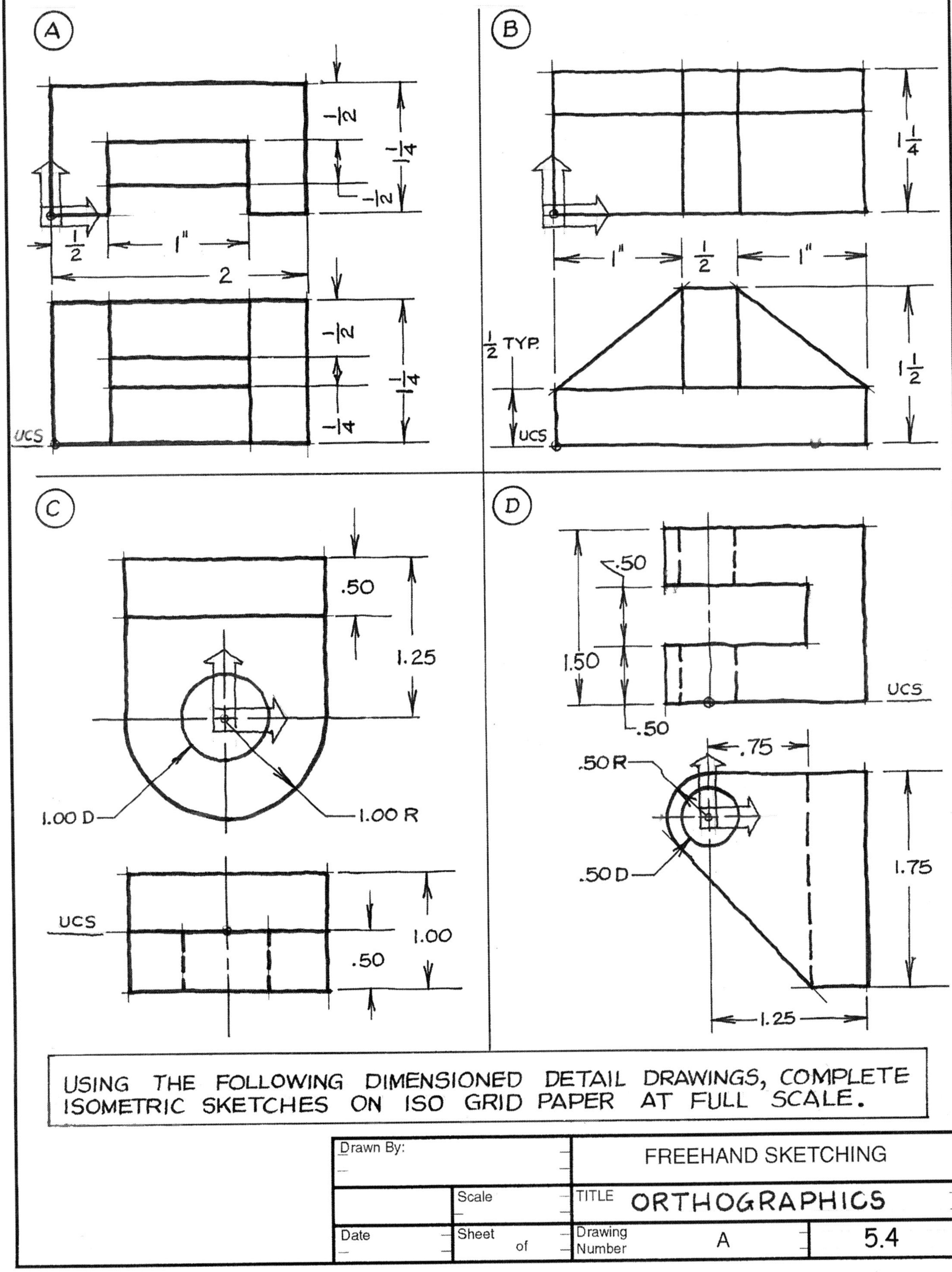

USING THE FOLLOWING DIMENSIONED DETAIL DRAWINGS, COMPLETE ISOMETRIC SKETCHES ON ISO GRID PAPER AT FULL SCALE.

Drawn By:		FREEHAND SKETCHING	
	Scale	TITLE ORTHOGRAPHICS	
Date	Sheet of	Drawing Number A	5.4

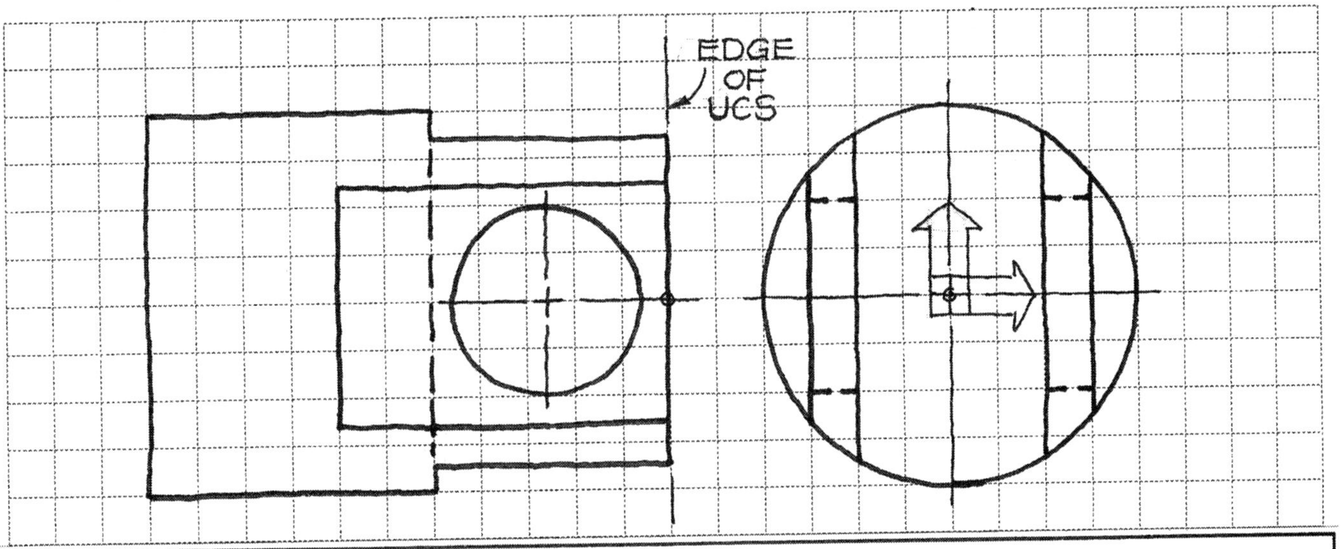

COMPLETE AN ISOMETRIC SKETCH OF THE FOLLOWING OBJECT AT FULL SCALE. CHOOSE THE MOST DESCRIPTIVE ORIENTATION.

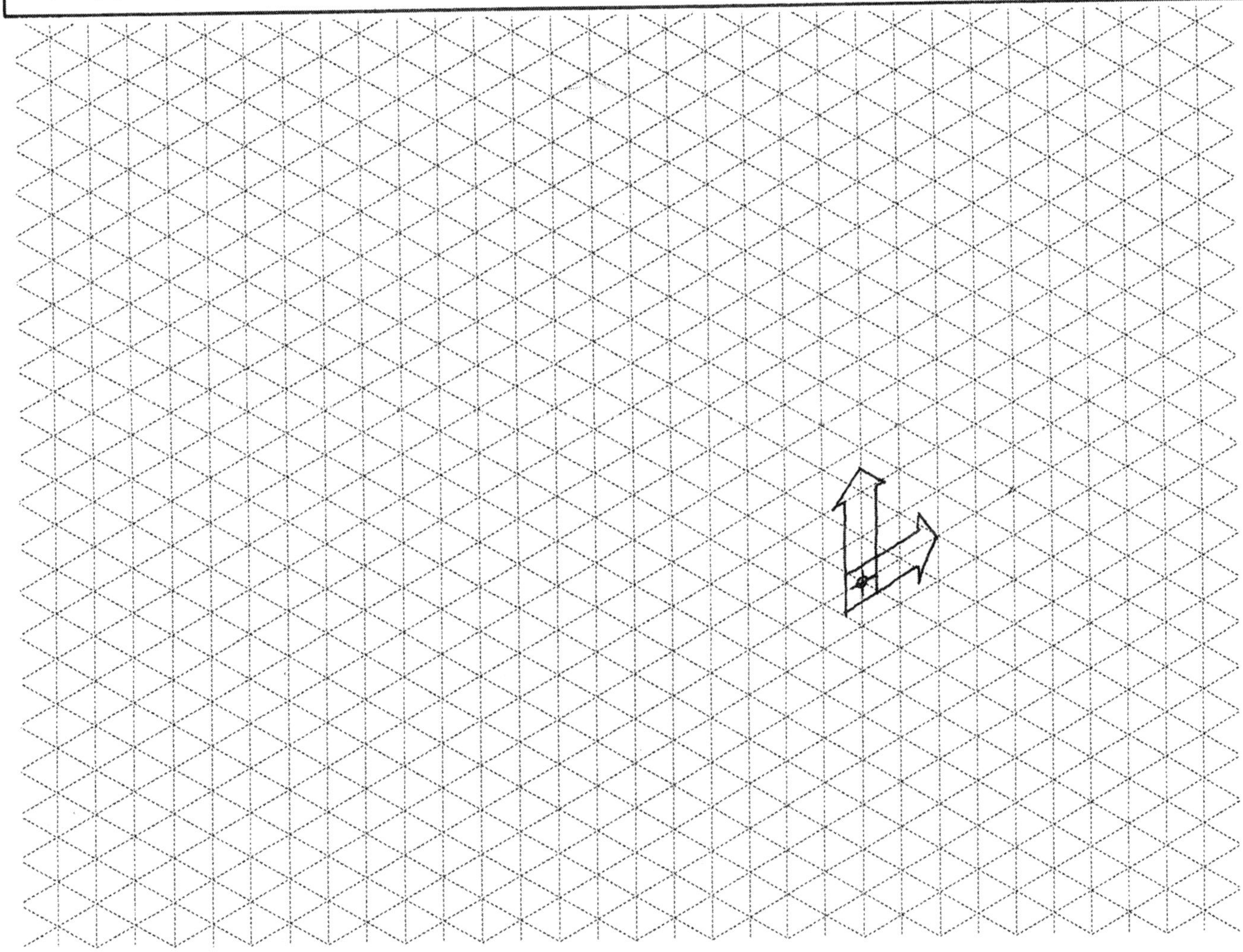

Drawn By:		FREEHAND SKETCHING	
	Scale	TITLE ISOMETRIC SKETCH	
Date	Sheet of	Drawing Number A	5.5

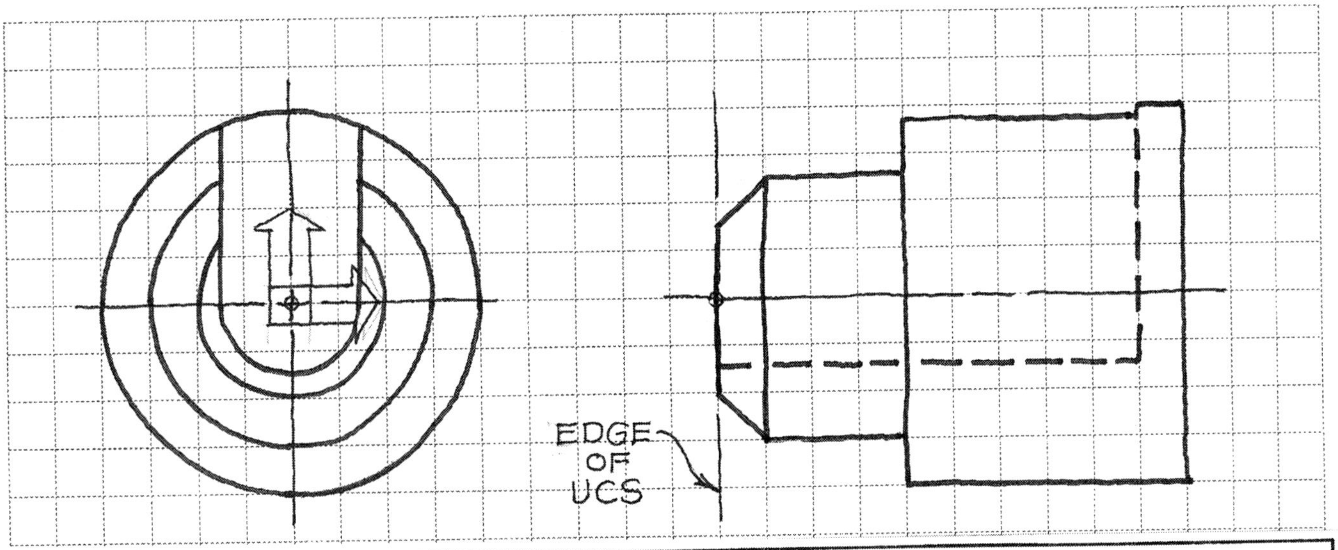

COMPLETE AN ISOMETRIC SKETCH OF THE FOLLOWING OBJECT AT FULL SCALE. CHOOSE THE MOST DESCRIPTIVE ORIENTATION.

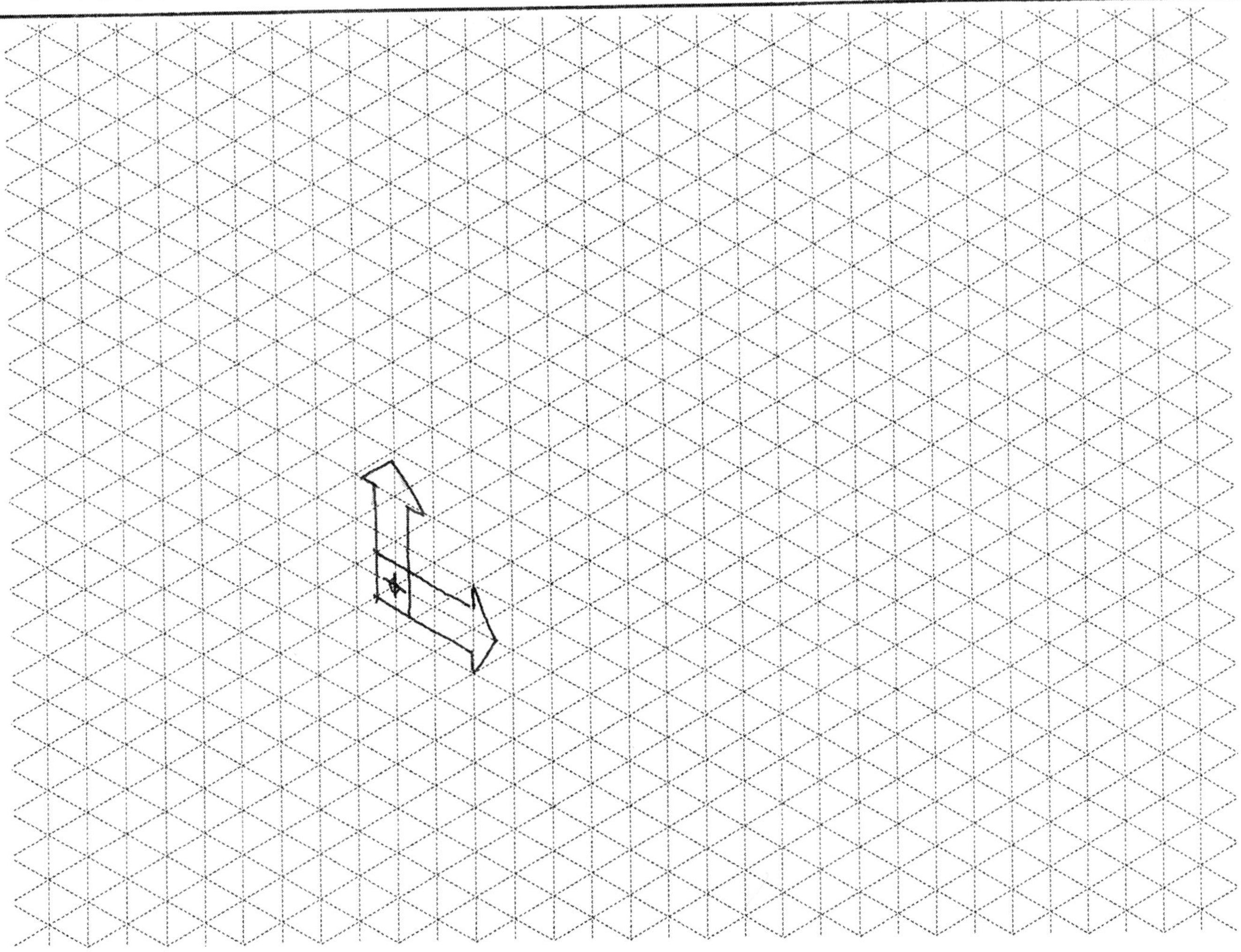

Drawn By:		FREEHAND SKETCHING	
	Scale	TITLE ISOMETRIC SKETCH	
Date	Sheet of	Drawing Number A	5.6

COMPLETE AN ISOMETRIC SKETCH OF THE FOLLOWING OBJECT AT THE SAME SCALE (1:1). USE ISO GRID PAPER.

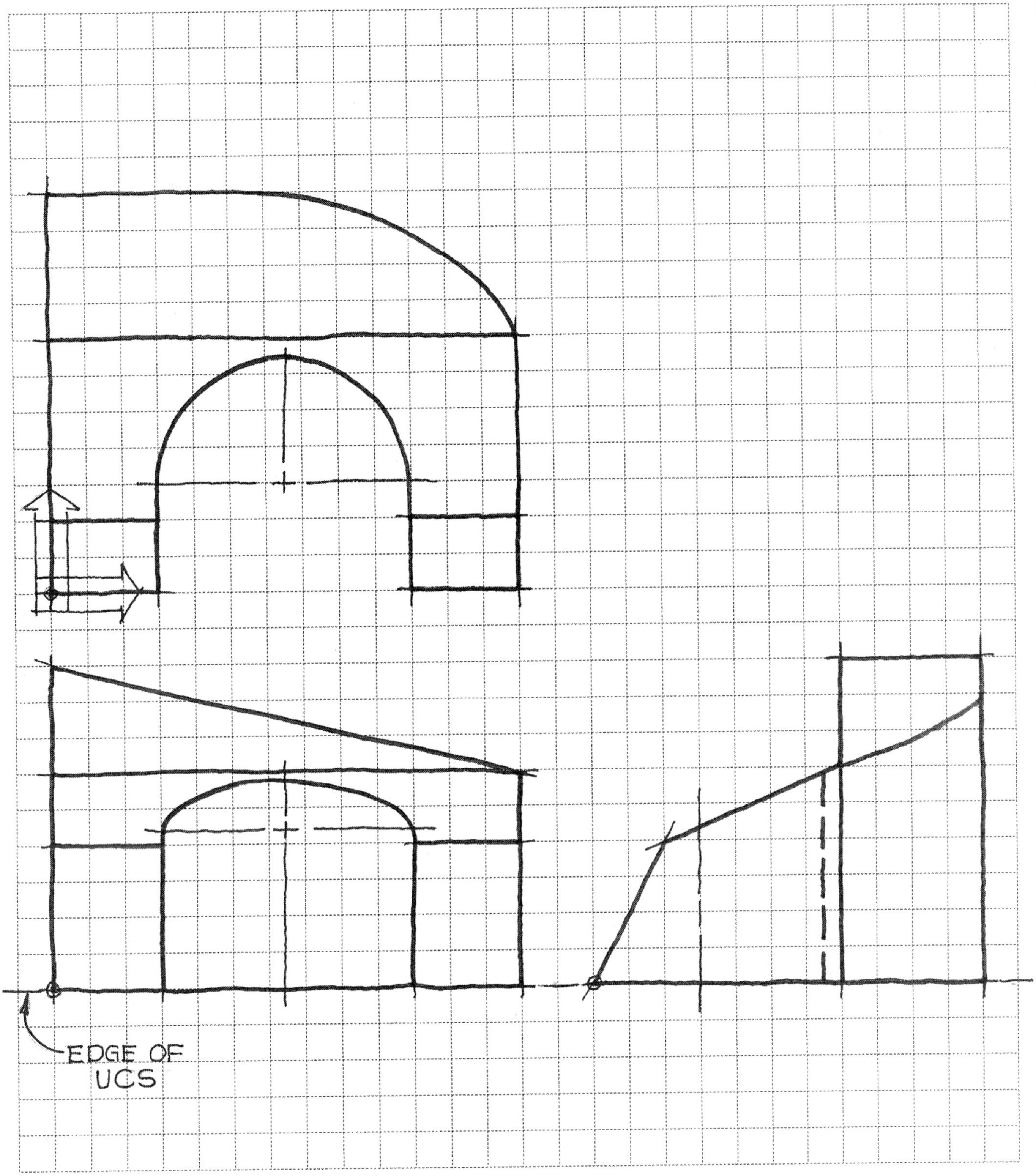

Drawn By:		FREEHAND SKETCHING	
	Scale	TITLE VIEWS FOR ISOMETRIC	
Date	Sheet of	Drawing Number A	5.7

6 Orthographic Reading

Coordinated with VMGED Chapter 10

Orthographic Reading—General Conditions

Orthographic views can be read much like a written language. When you "read" a view, you interpret lines to form a mental picture of what the object looks like. To fully understand an object's geometry from only one view is difficult. When two or more orthogonal views are arranged in close proximity (aligned), you can glance from one view to the next, mentally noting a feature's position and shape, and in doing so, become familiar with its shape. This process is called *reading the views*.

Understanding the relationship of views and positions in space is important. In Fig. 6-1, note the position of the block on top of the object in each view. To be able to read the views, you need to be able to communciate where you are in space and in which way you are moving. Can you verbalize how the various views were achieved? Sketch a vector on the pictorial indicating the direction of each view.

Reading is an activity that makes you a better sketcher and visualizer. Interestingly, the better pictorial sketcher you are, the better you can solve reading problems. It was for that reason pictorial sketching was presented first.

Understanding Surfaces

The position of a surface determines how it will appear in orthographic views. In general, a plane can appear in true shape, as an edge (a line), or in an infinite number of foreshortened positions. There are two unchangeable truths about planes:

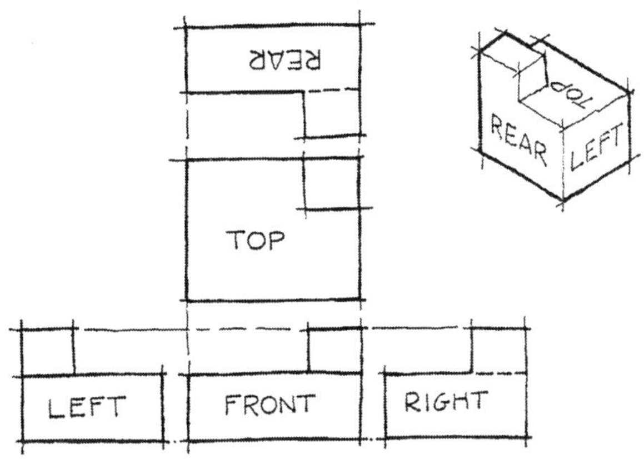

Fig. 6-1. Spatial interpretation of two-dimensional views.

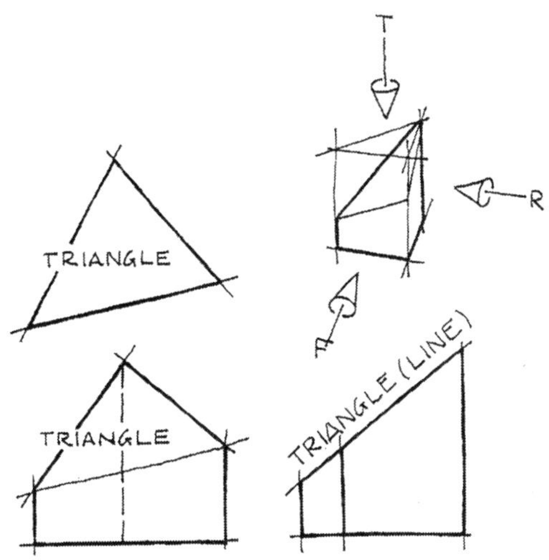

Fig. 6-2. Basic shape of a plane can't be changed.

Truth 1. The basic shape of a plane doesn't change from view to view. A circular plane appears as a circle, ellipse, or line; a square plane appears as a square, rhombus, or line; a triangular plane appears as a triangle, a smaller (squeezed) triangle, or a line (Fig. 6-2, previous page).

Truth 2. The height, width, and depth of a plane relative to a stable axis system doesn't change. If a plane is 2" above the base in the front view, it will be 2" above the base in the side view because both views show height measurements (Fig. 6-3).

Sheet 6.0 (page 107) presents a pictorial view and top, front, and right side views of an object with many surfaces. Note how much easier it is to visualize the orthographic views when the pictorial is present.

A plane can be perpendicular to one of the directions of sight for a principal view—top, front, or right side. These planes are called *horizontal*, *vertical*, and *profile*. If a plane is parallel to only one direction, it is said to be *inclined*. If a plane is at an angle to all three principal directions of sight, it is said to be *oblique*.

Assignment: Using the pictorial on sheet 6.0 as a guide, assign the correct labels to the planes in the top, front, and right side views. Use arrows when necessary to point to corners. Using the definitions above, complete boxes A-L in appropriate technical lettering.

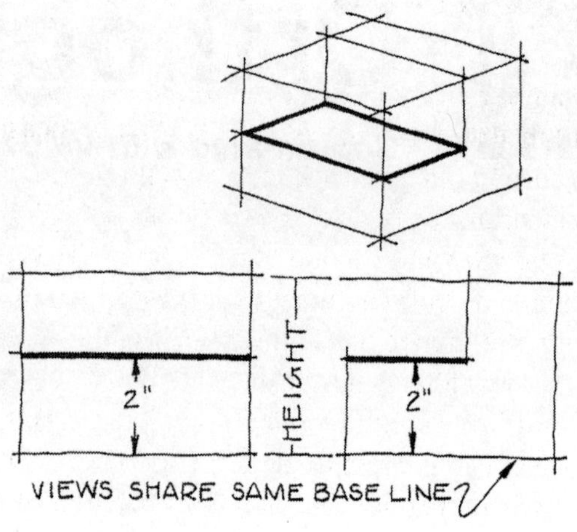

Fig. 6-3. Position in space can't be changed.

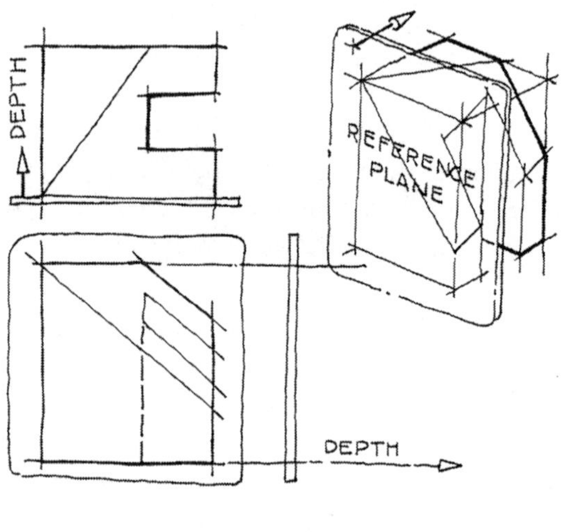

Fig. 6-4. Use of reference plane. See also sheet 6.8.

Reading Exercises

When you read views, you generally are either trying to find information already in the view or complete a new view. A technique that helps you find information is the use of *reference planes*. Figure 6-4 shows an object in the front and top views. Assume you want to sketch the right side view. Note that a plane of reference has been installed at the front of the object in the top view, controlling depth measurements. The same reference plane appears as an edge in the side view. Depth measurements in the side view will be identical to their corresponding depth measurements in the top view. Height information is obtained from the front view.

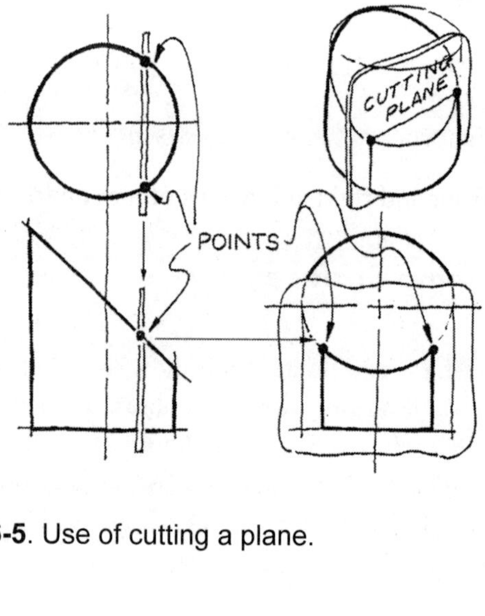

Fig. 6-5. Use of cutting a plane.

Use of Cutting Planes

Similar to a reference plane is a *cutting plane*. The difference lies in its use. A reference plane is used to measure, whereas a cutting plane is used to reveal intersections. These intersections provide common points so that geometry can be drawn. A cutting plane is normally used to locate points on curved intersections, as shown in Fig. 6-5. As with a reference plane, the cutting plane should be passed perpendicular (normal) to some line of sight. Figure 6-6 shows two cutting planes that have been passed in the front view to locate the same points. The horizontal cutting plane is perpendicular to the top direction of sight. The vertical cutting plane is perpendicular to the side view line of sight. Either one will work.

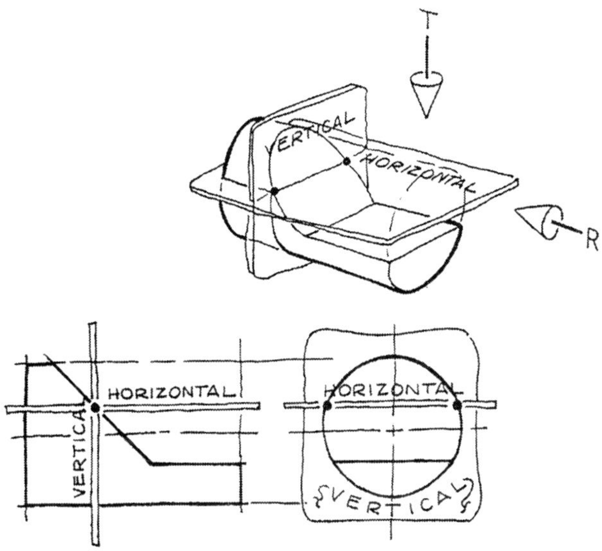

Fig. 6-6. Position of reference normal to line of sight.

Practical Example

In Fig. 6-7, front and right side views of a cylindrical object are shown. You are required to create a top view.

Step 1. Sketch a general shape for the new view that describes the overall width (from the front view) and depth (from the side view).

Step 2. Pass a plane of reference at a convenient location vertically through the cylinder in the side view. This appears as a horizontal line in the top view.

Step 3. In the side view, locate the general feature depth perpendicular to this reference.

Step 3.1. Find the corresponding points in the front view. Transfer the depth of these points (in the top view) along the projections from the front view.

Step 3.2. Locate the limits (farthest front, rear, top, and bottom) of the curve in the side view. Transfer these limits to the front view and project to the top view.

Step 3.3. Transfer the depth measurements (from the reference plane) for these limits from the side view to the top view.

Step 3.4. Pass cutting planes in the side view (as necessary) to locate intermediate points along the curve between the limits.

Step 3.5. Find these intermediate points in the front view.

Step 3.6. If needed, locate cutting planes in the top view by measuring their distances from the reference plane. Measure depth in side view and transfer to top view.

Step 4. Complete the curve and then complete the entire view.

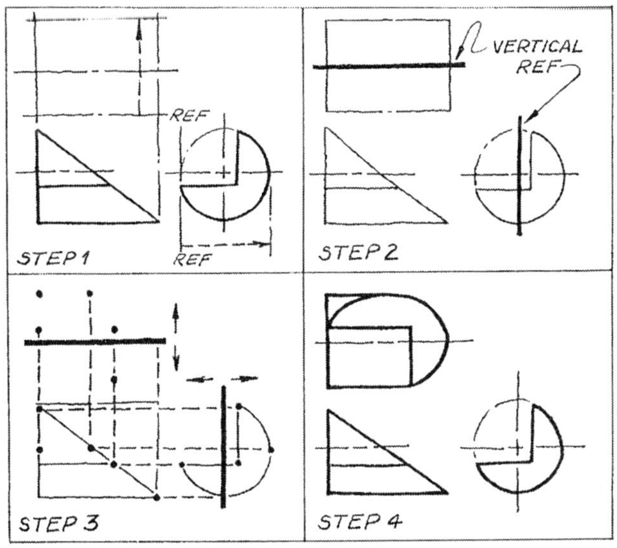

Fig. 6-7. Solution of orthographic views.

CAD Example—2D

CAD systems automate the two-dimensional constructions previously described. Because software will report distances very accurately, and perpendiculars can be automatically constructed, new views can easily be constructed using reference and cutting planes.

CAD Example—3D

With a three-dimensional model, *any* view is possible. The technique similar to a reference plane is often called a *construction plane*. A construction plane may be set at any position and at any orientation in space. A viewing position taken perpendicular to a construction plane results in a view of the geometry as if projected onto that plane. Figure 6-8 shows a series of construction planes and the UCS axes for views constructed on those planes.

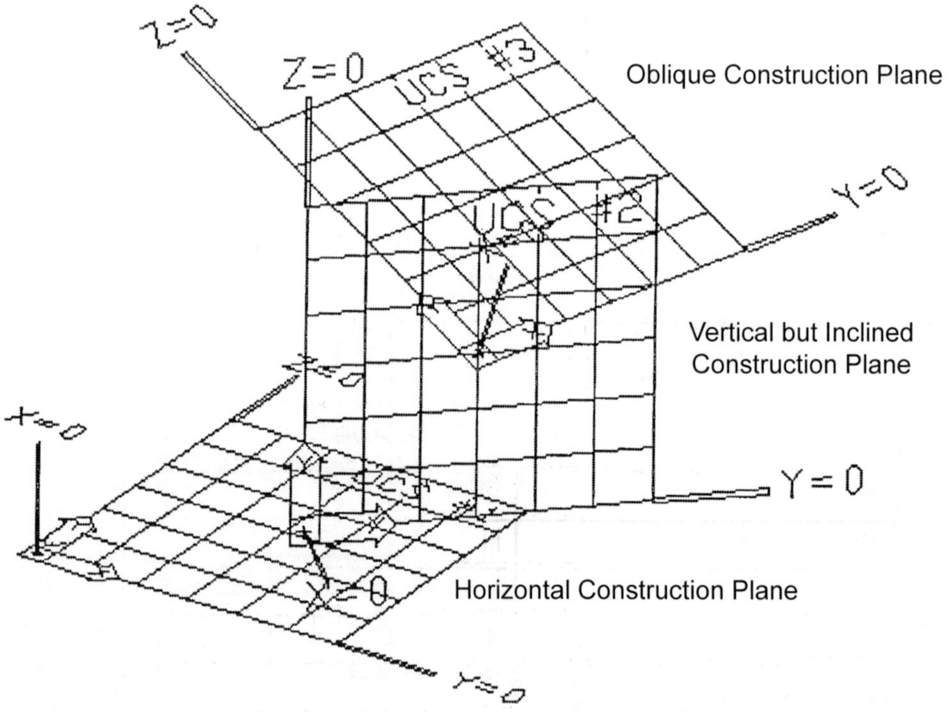

Fig. 6-8. Views in 3D CAD generated by placement of construction planes.

1. Label the surfaces and edges on the orthographic multiviews to match the letters shown on the isometric pictorial.

2. Complete the table below by identifying each type of surface. Shade ALL surfaces on the multiview and the isometric pictorial drawing as follows.

SURFACE Identification and Definition	*Color*
FRONTAL: A normal surface or plane parallel to the front of an object.	Red
HORIZONTAL: A normal surface or plane parallel to the top of an object.	Green
PROFILE: A normal surface or plane parallel to the right side of an object.	Blue
INCLINED: A surface or plane sloped at one angle so that it appears as foreshortened in two of the principal views of an object and as an edge in the third view.	Yellow
OBLIQUE: A surface or plane sloped at two angles so that it appears as foreshortened in all three principal views of an object.	Orange

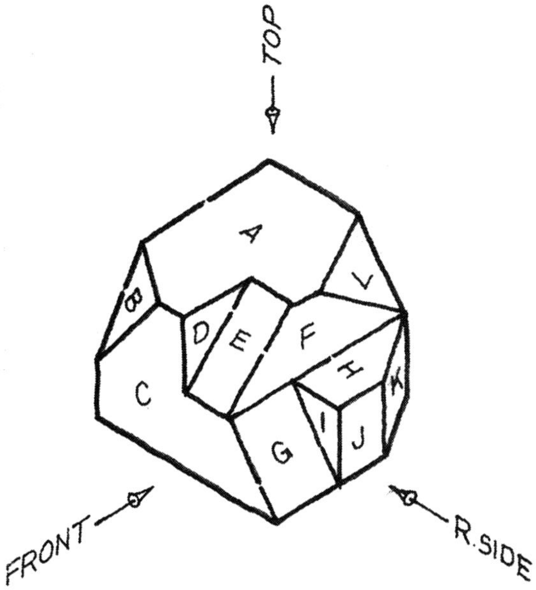

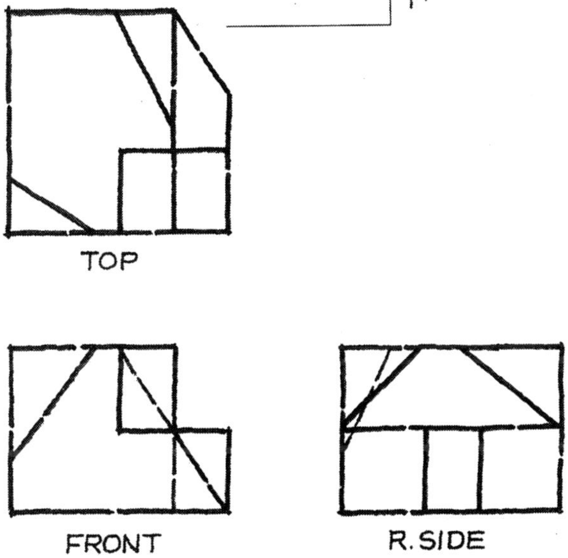

A		G	
B		H	
C		I	
D		J	
E		K	
F		L	

Drawn By:		FREEHAND SKETCHING	
	Scale	TITLE READING SURFACES	
Date	Sheet of	Drawing Number A	6.0

IN THE TABLES, LIST THE NUMBERS CORRESPONDING TO THE LETTERS.

	Top	Front	Side
EXAMPLE → A	21	6	14
B			
C			
D			
E			
F			
G			

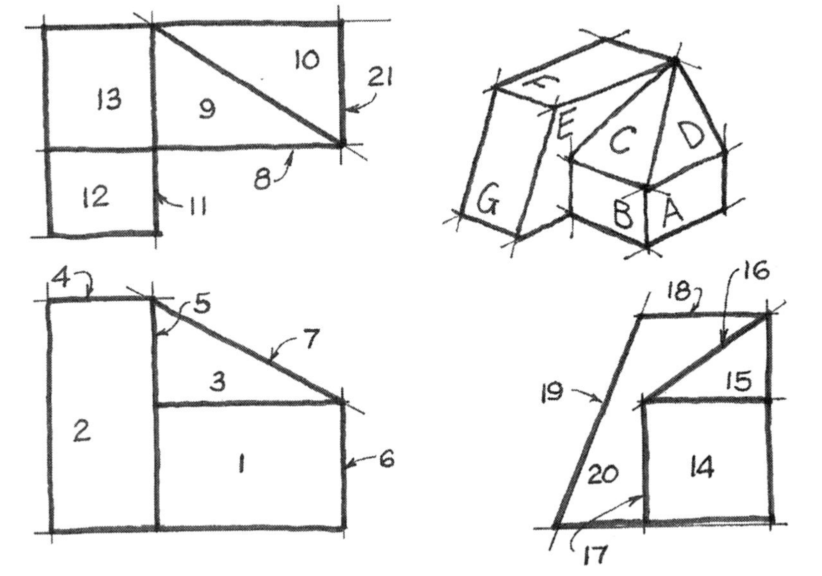

	Top	Front	Side
A			
B			
C			
D			
E			
F			
G			
H			
I			

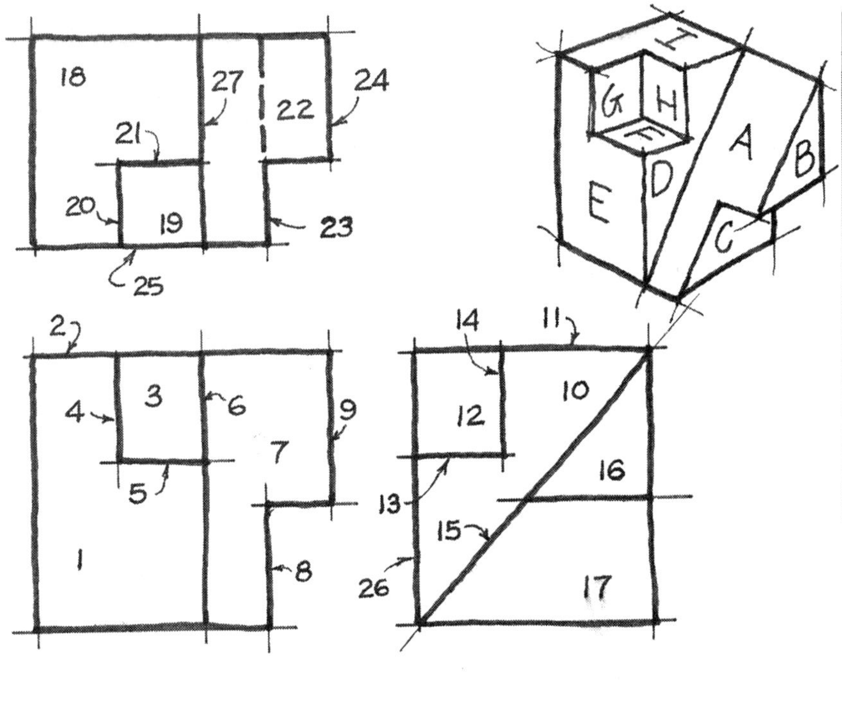

Drawn By:		FREEHAND SKETCHING	
	Scale	TITLE READING	
Date	Sheet of	Drawing Number A	6.1

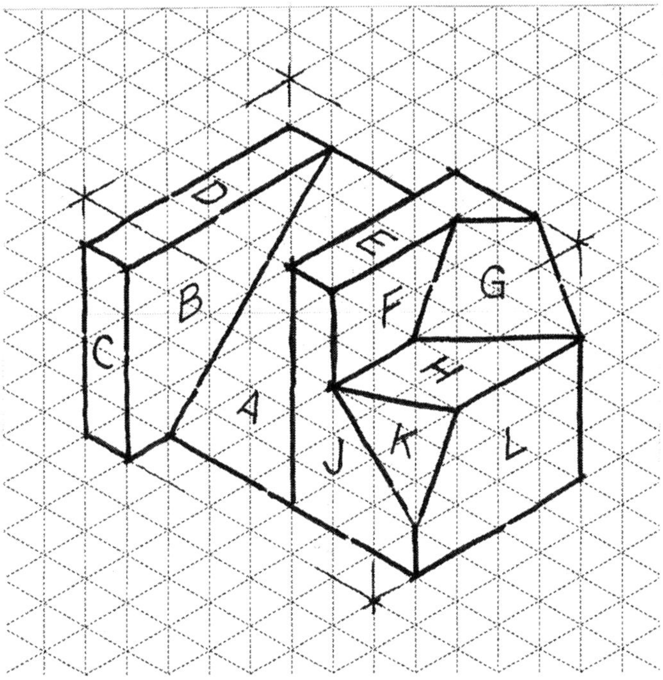

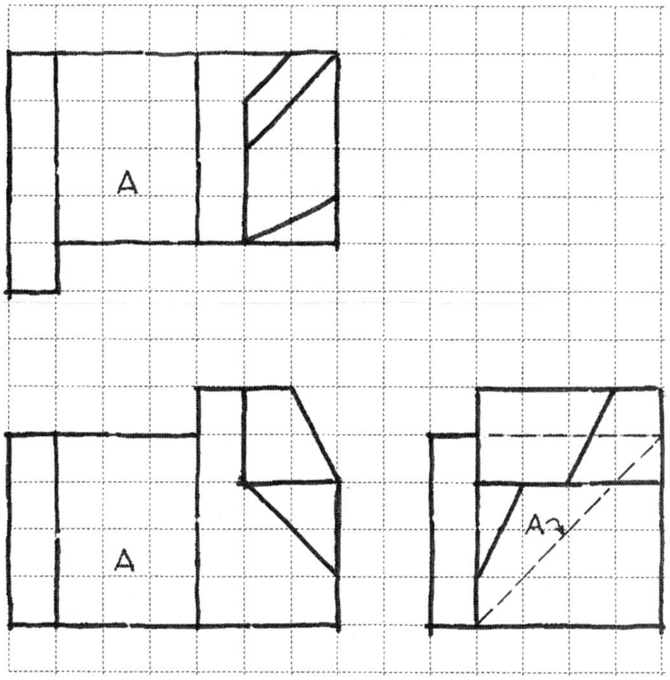

SURFACE IDENTIFICATION: Using the following descriptions, complete the table as shown in the example.

1 = TRUE SHAPE: When a surface is perpendicular to your line of sight in a given orthographic view, the surface is referred to as true shape.

2 = LINE: In an orthographic view, the edge view of a surface is represented by a line. The direction of the surface is therefore parallel to your line of sight in that view.

3 = FORESHORTENED: When a surface is sloped at an angle so that it is neither parallel nor perpendicular to your line of sight in a principal view, the surface is referred to as foreshortened.

	ISO	TOP	FRONT	R.SIDE
EXAMPLE	A	3	3	2
	B			
	C			
	D			
	E			
	F			
	G			
	H			
	J			
	K			
	L			

Drawn By:		FREEHAND SKETCHING	
	Scale	TITLE READING	
Date	Sheet of	Drawing Number A	6.2

COMPLETE THE MISSING VIEW OF EACH SINGLE SOLID OBJECT.

Drawn By:		FREEHAND SKETCHING	
	Scale	TITLE MISSING VIEWS	
Date	Sheet of	Drawing Number A	6.3

COMPLETE THE MISSING VIEW OF EACH SINGLE SOLID OBJECT.

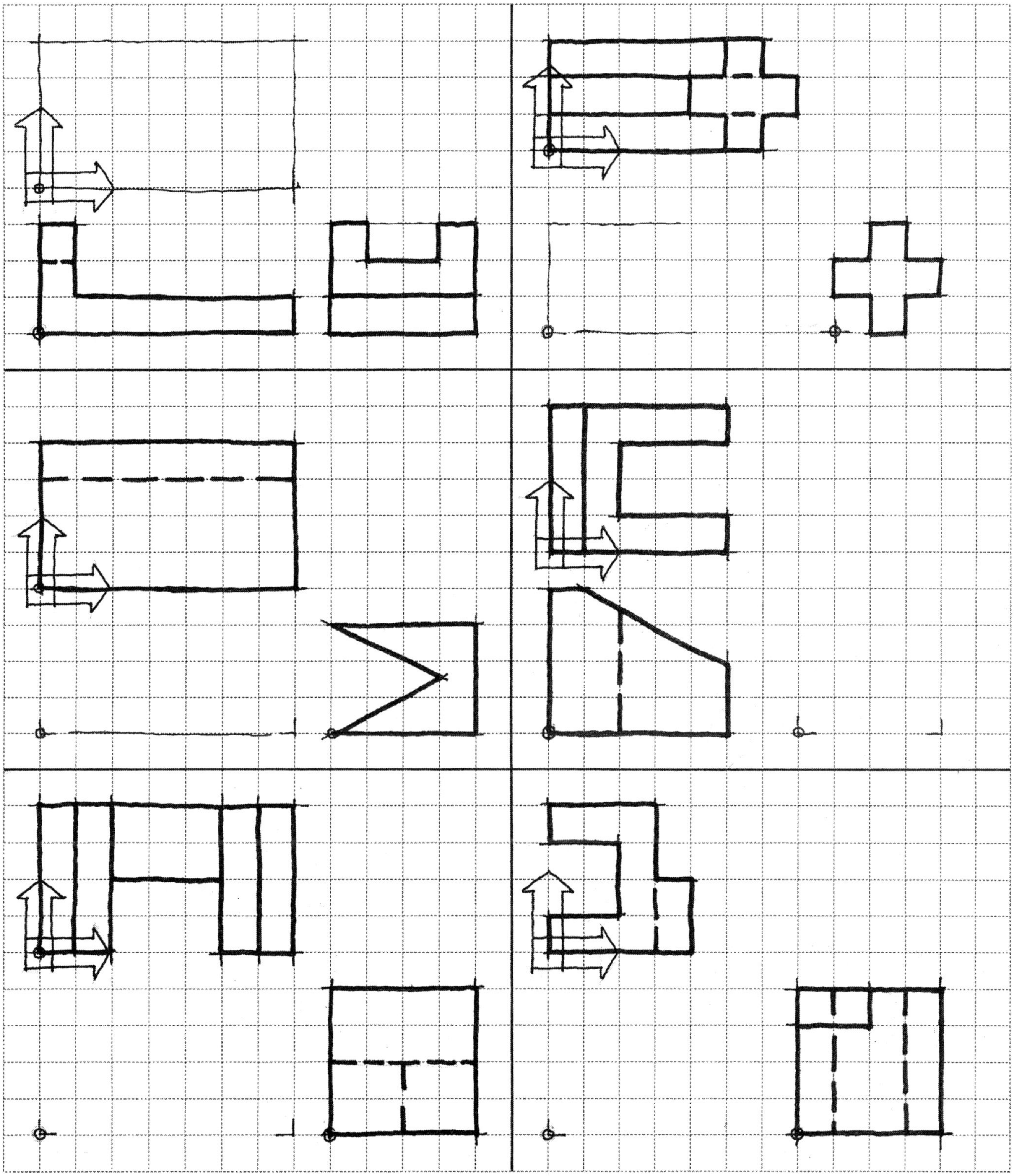

Drawn By:		FREEHAND SKETCHING	
	Scale	TITLE MISSING VIEWS	
Date	Sheet of	Drawing Number A	6.4

COMPLETE THE MISSING VIEW OF EACH SINGLE SOLID OBJECT.

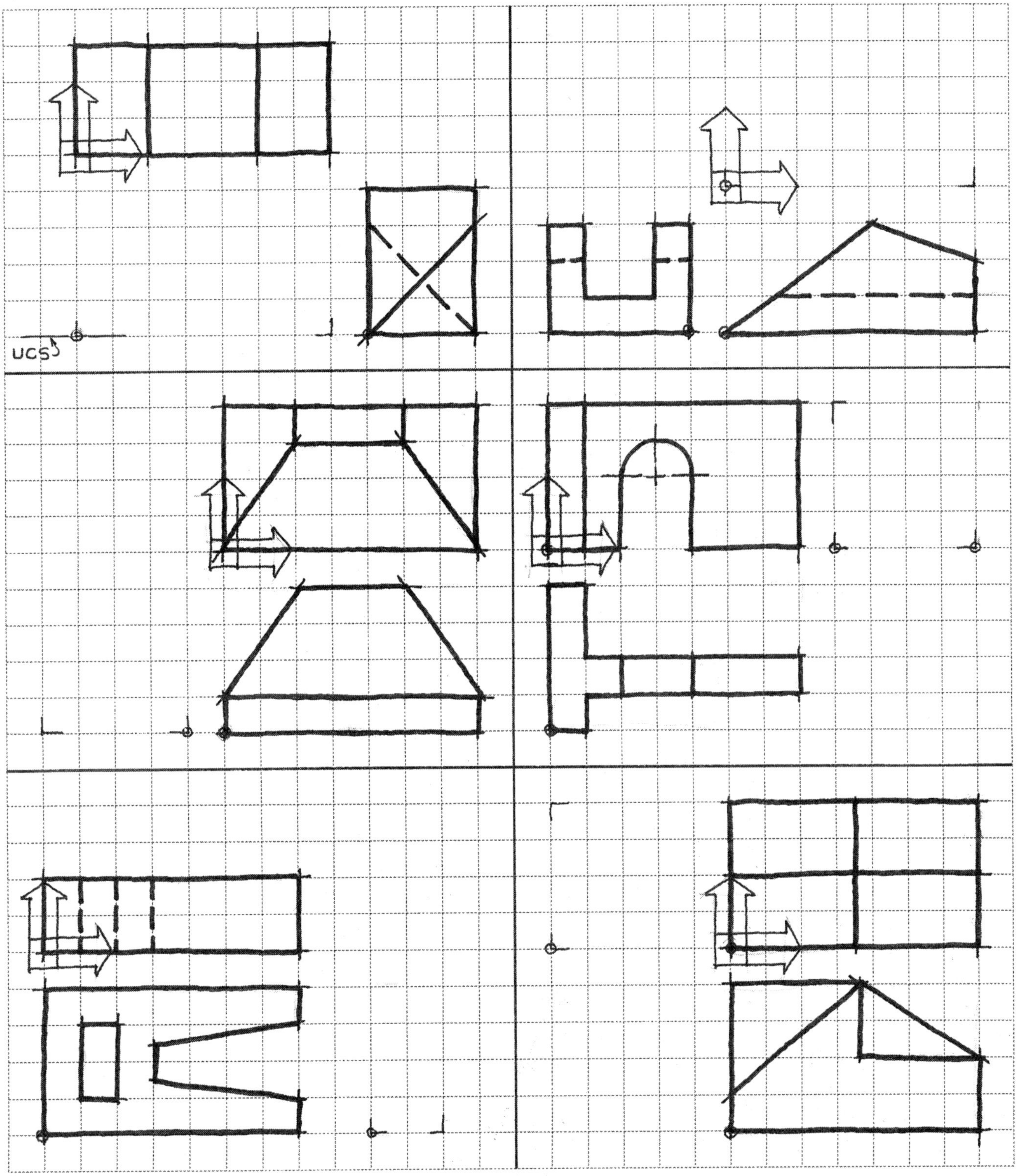

Drawn By:		FREEHAND SKETCHING	
	Scale	TITLE MISSING VIEWS	
Date	Sheet of	Drawing Number A	6.5

COMPLETE THE MISSING VIEW OF EACH SINGLE SOLID OBJECT.

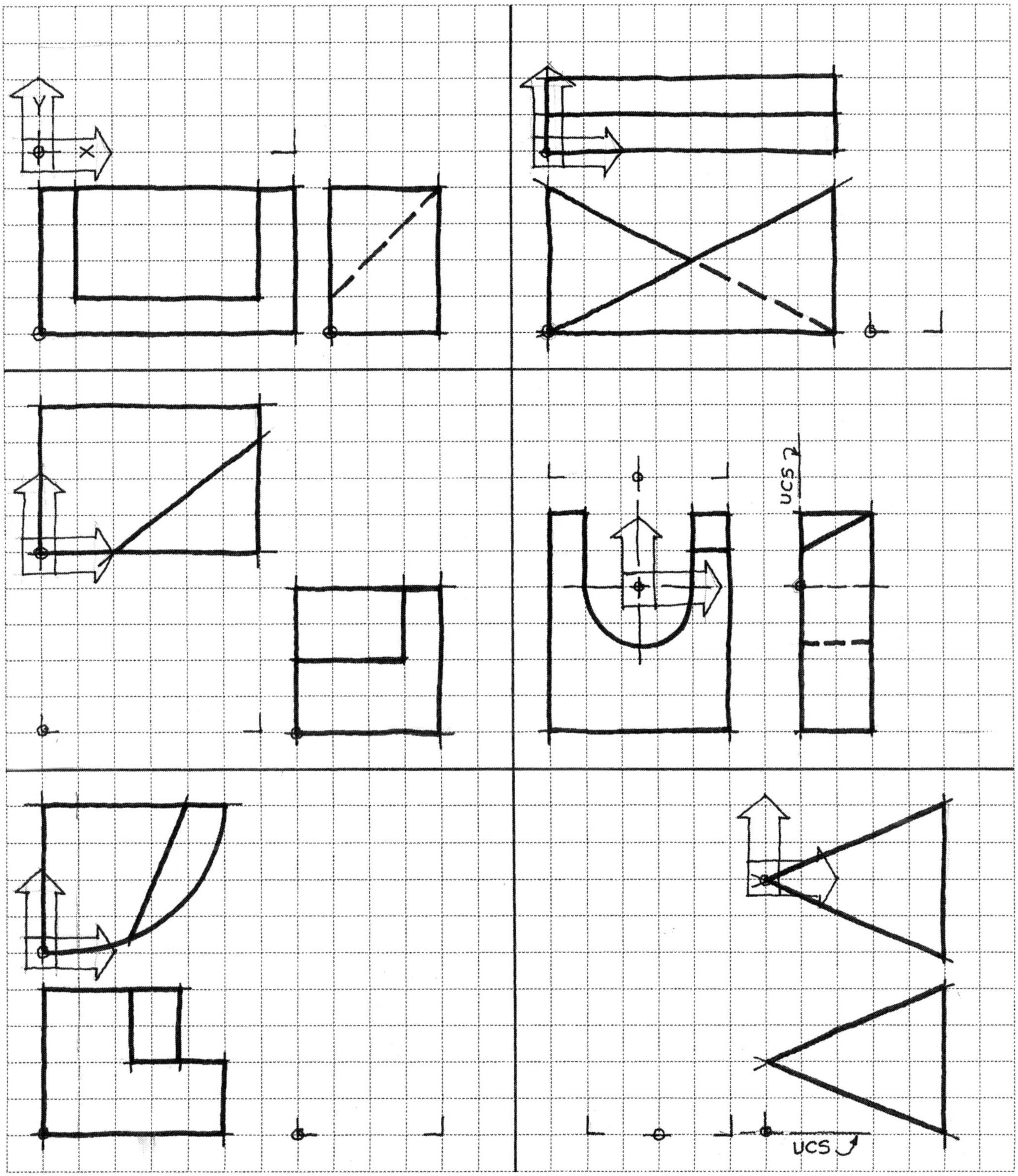

Drawn By:		FREEHAND SKETCHING	
	Scale	TITLE MISSING VIEWS	
Date	Sheet of	Drawing Number A	6.6

COMPLETE THE MISSING VIEW OF EACH SINGLE SOLID OBJECT.

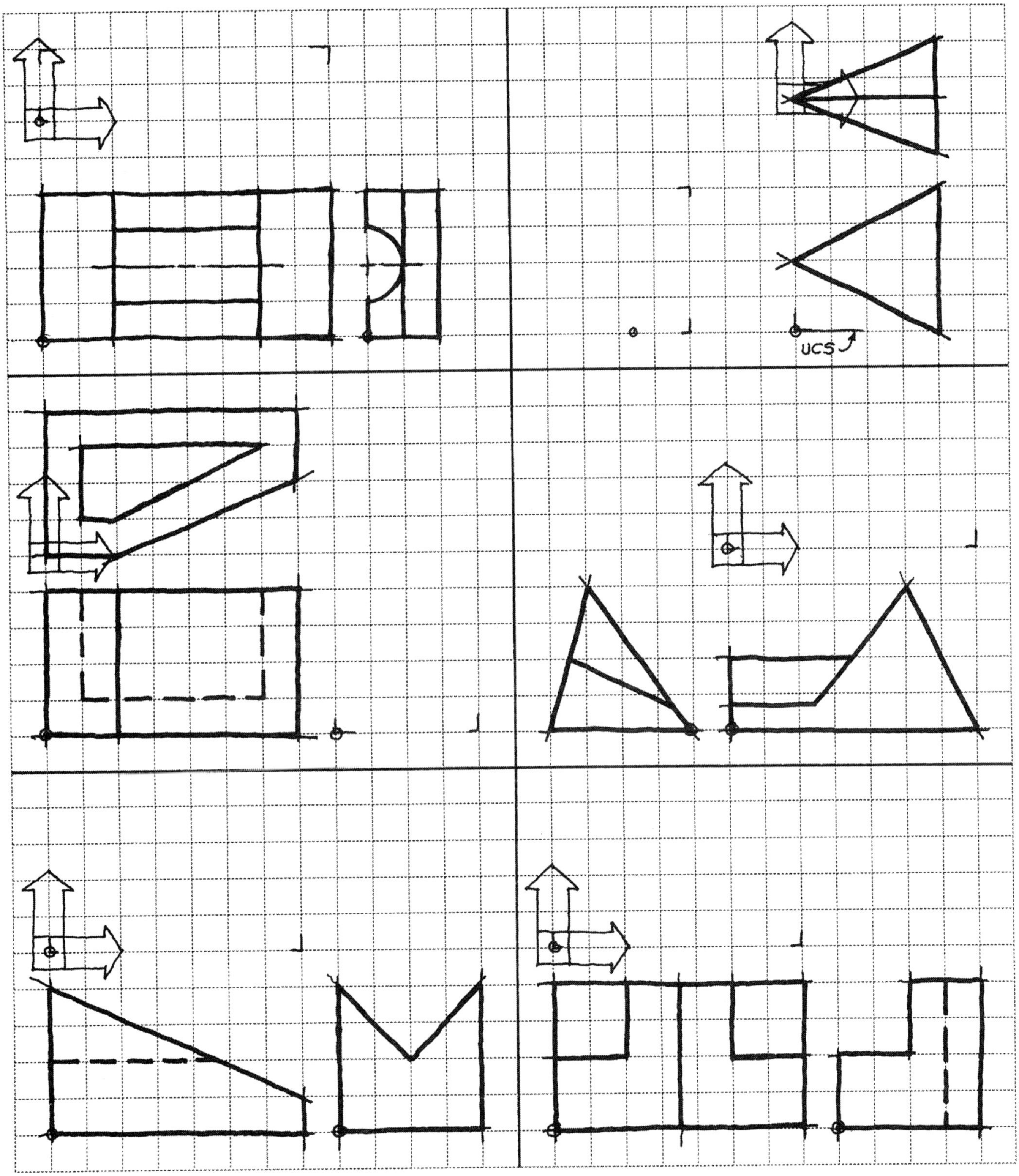

Drawn By:		FREEHAND SKETCHING	
	Scale	TITLE MISSING VIEWS	
Date	Sheet of	Drawing Number A	6.7

UCS – USER COORDINATE SYSTEM AS A REFERENCE PLANE

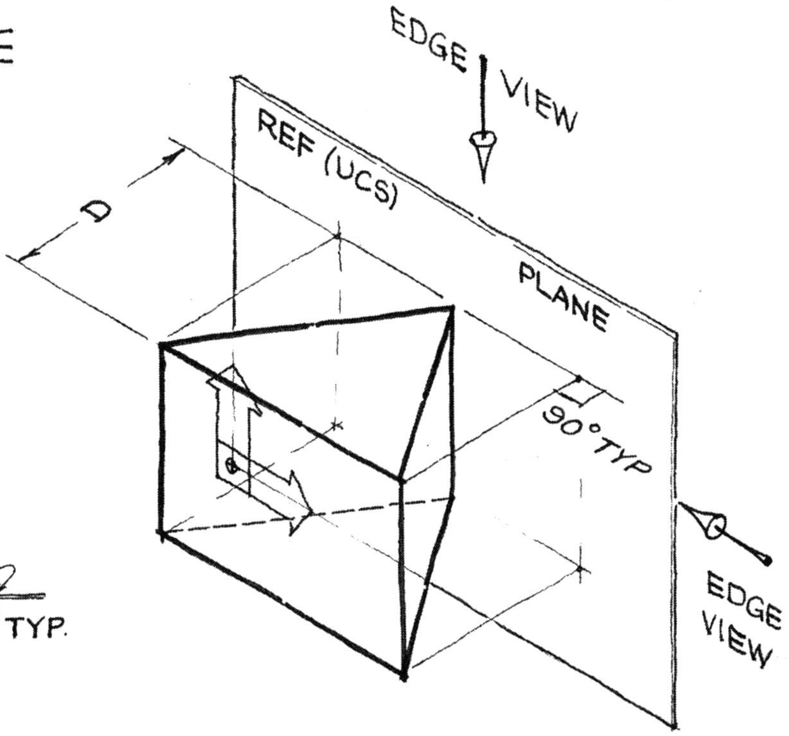

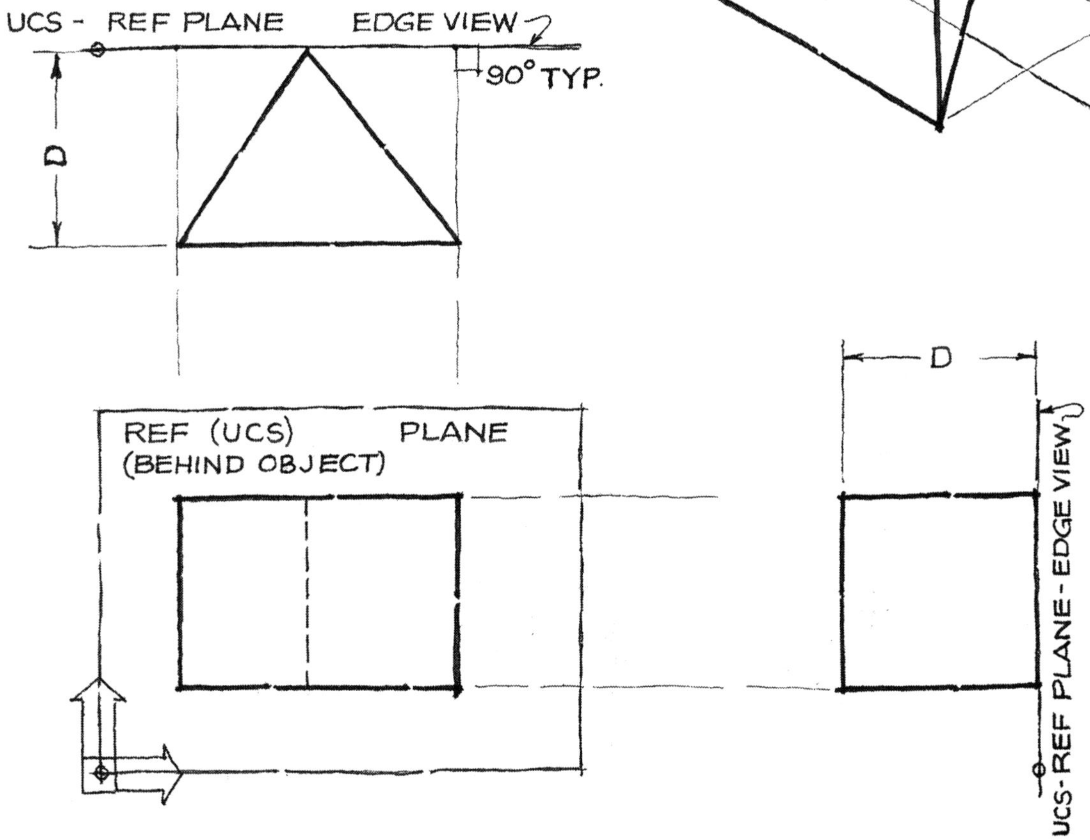

Drawn By:		FREEHAND SKETCHING	
	Scale	TITLE REFERENCE PLANES	
Date	Sheet of	Drawing Number A	6.8

USING A REFERENCE PLANE AS SHOWN IN PROBLEM (A), SOLVE THE MISSING VIEWS IN ALL OF THE PROBLEMS BELOW.

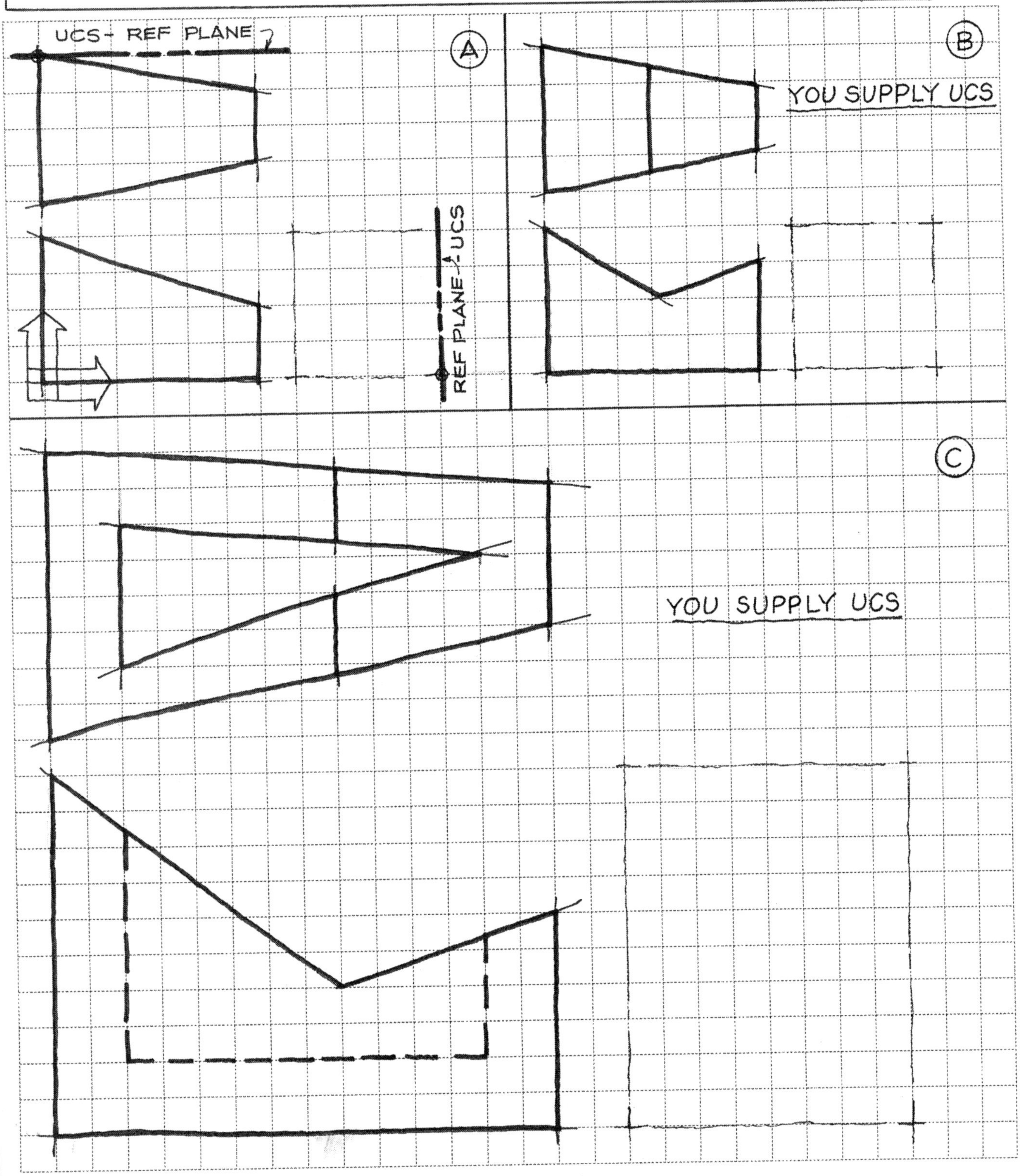

Drawn By:		FREEHAND SKETCHING	
	Scale	TITLE REFERENCE PLANES	
Date	Sheet of	Drawing Number A	6.9

COMPLETE THE TOP VIEW

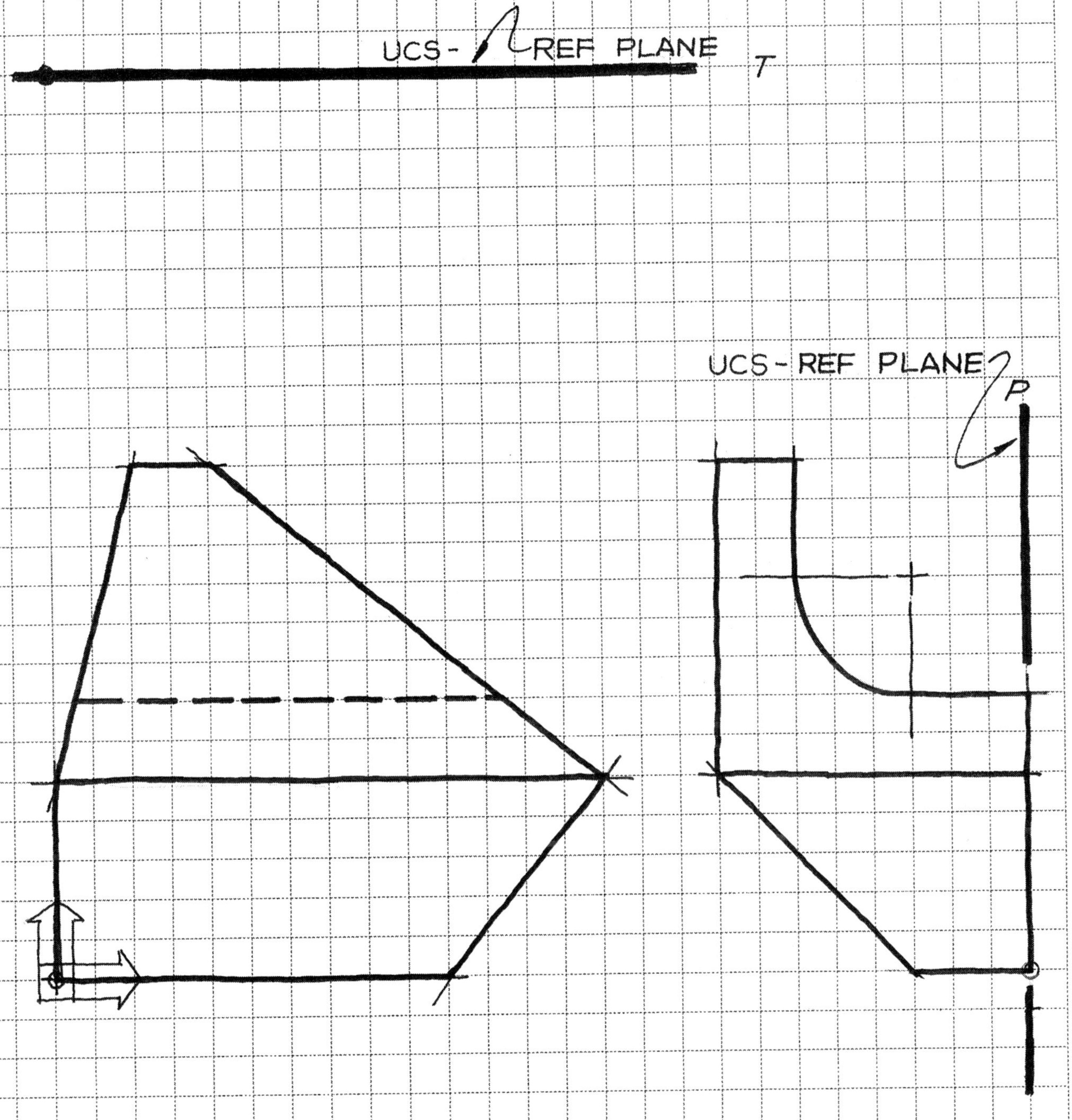

Drawn By:		FREEHAND SKETCHING	
	Scale	TITLE REFERENCE PLANES	
Date	Sheet of	Drawing Number A	6.10

SUPPLY THE MISSING VIEWS. THE VIEWS THAT ARE GIVEN ARE CORRECT AND COMPLETE.

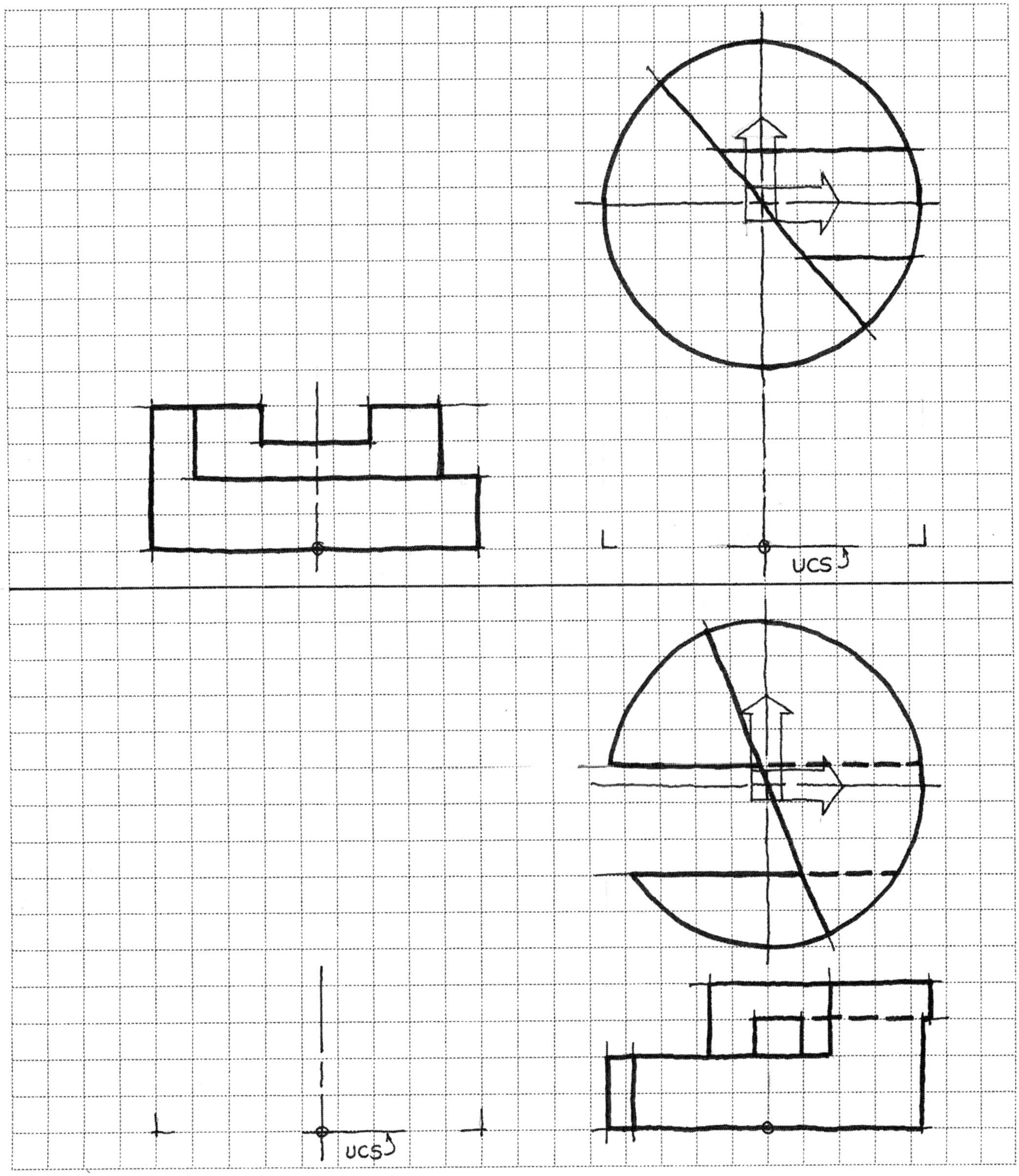

Drawn By:		FREEHAND SKETCHING	
	Scale	TITLE REFERENCE PLANES	
Date	Sheet of	Drawing Number A	6.11

SUPPLY THE MISSING VIEWS. THE VIEWS THAT ARE GIVEN ARE CORRECT AND COMPLETE. USE CUTTING PLANES.

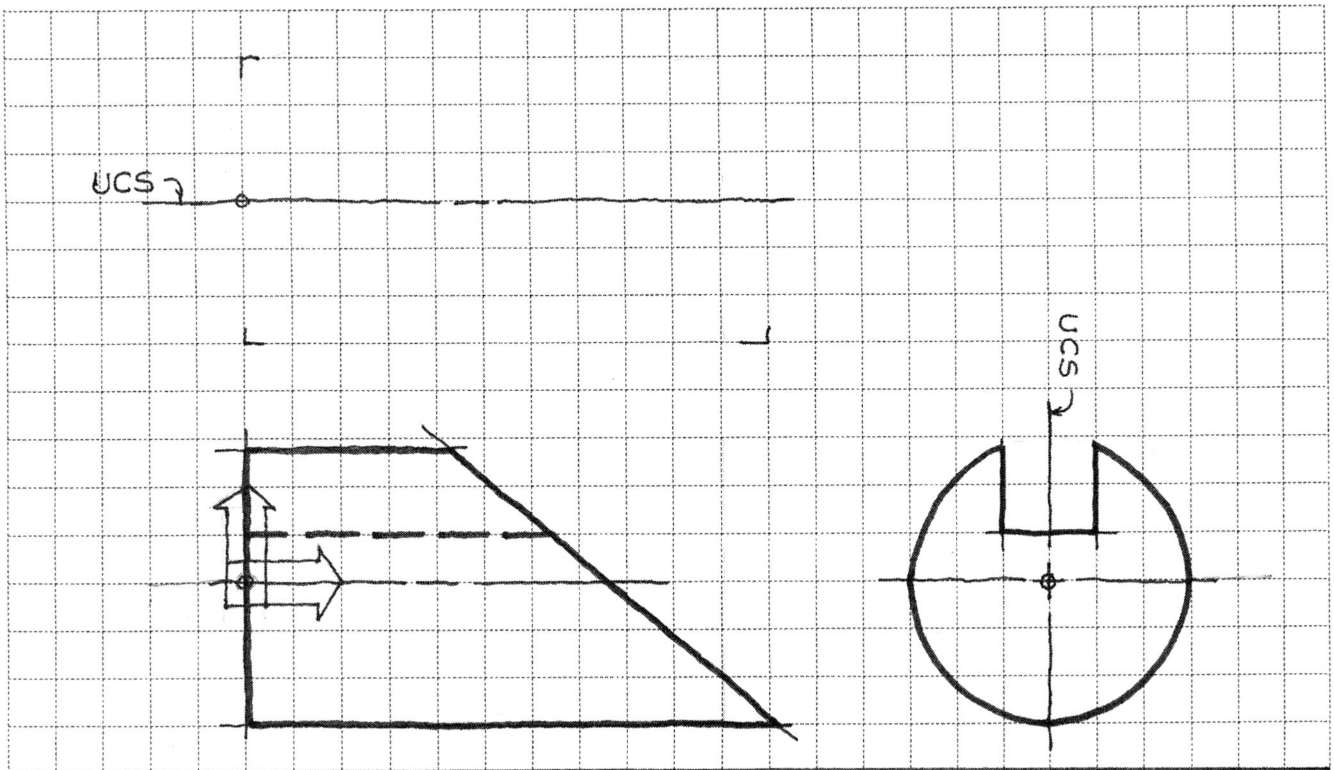

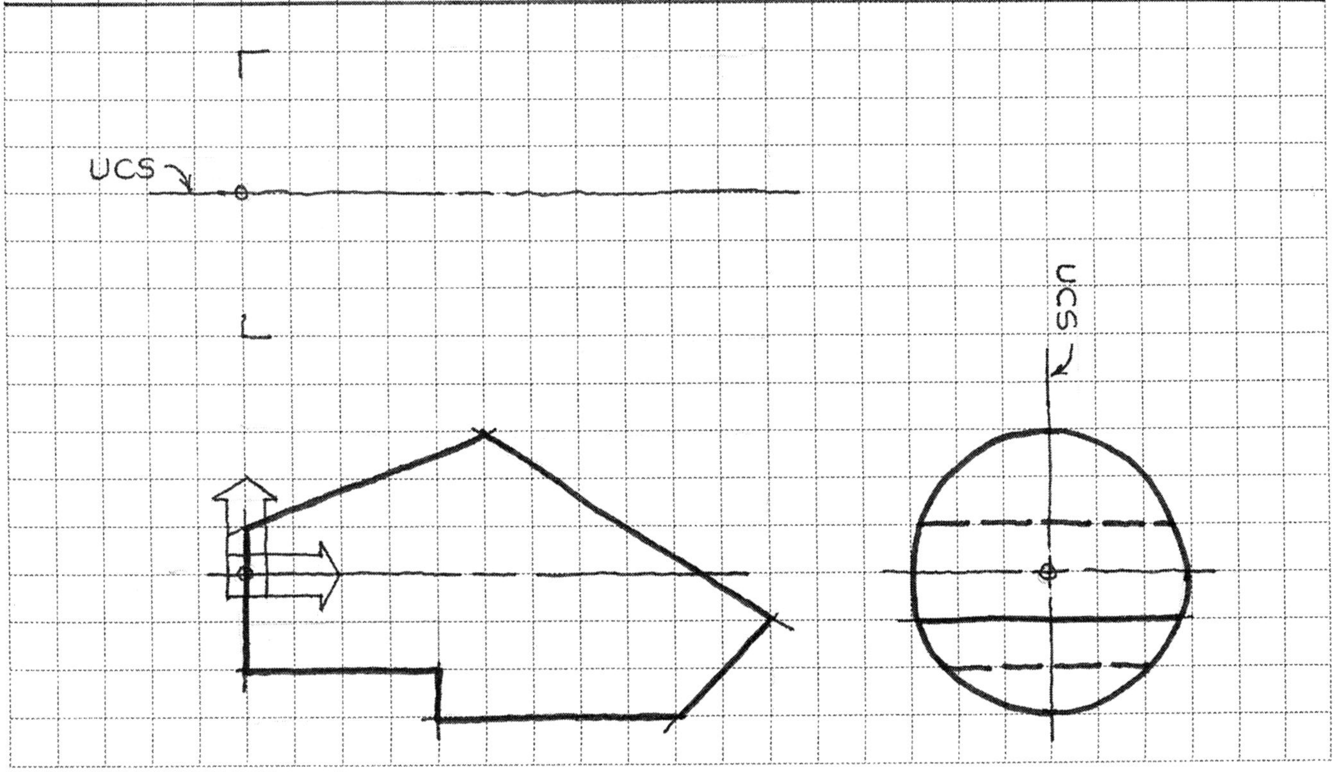

Drawn By:		FREEHAND SKETCHING	
	Scale	TITLE CUTTING PLANES	
Date	Sheet of	Drawing Number A	6.12

SUPPLY THE MISSING VIEW. THE VIEWS THAT ARE GIVEN ARE CORRECT AND COMPLETE. USE CUTTING PLANES.

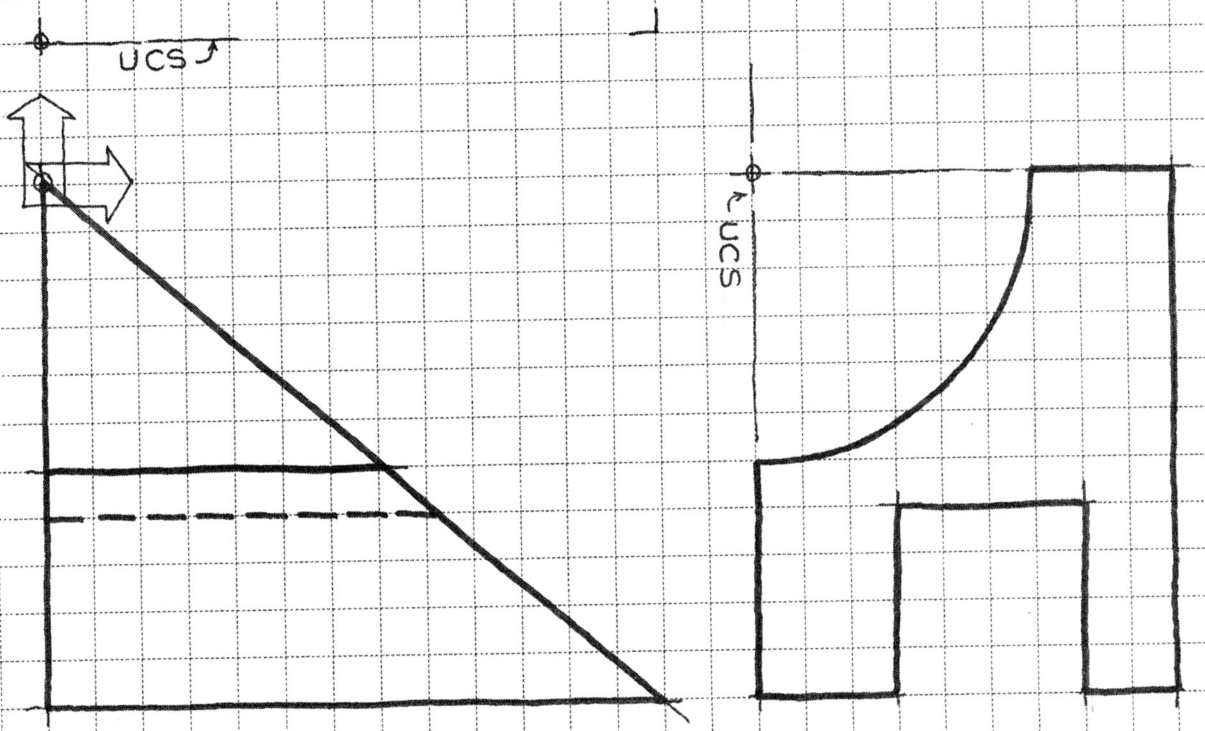

Drawn By:		FREEHAND SKETCHING	
	Scale	TITLE CUTTING PLANES	
Date	Sheet of	Drawing Number A	6.13

SUPPLY THE MISSING VIEW. THE VIEWS THAT ARE GIVEN ARE CORRECT AND COMPLETE. (MULTIPLE SOLUTIONS)

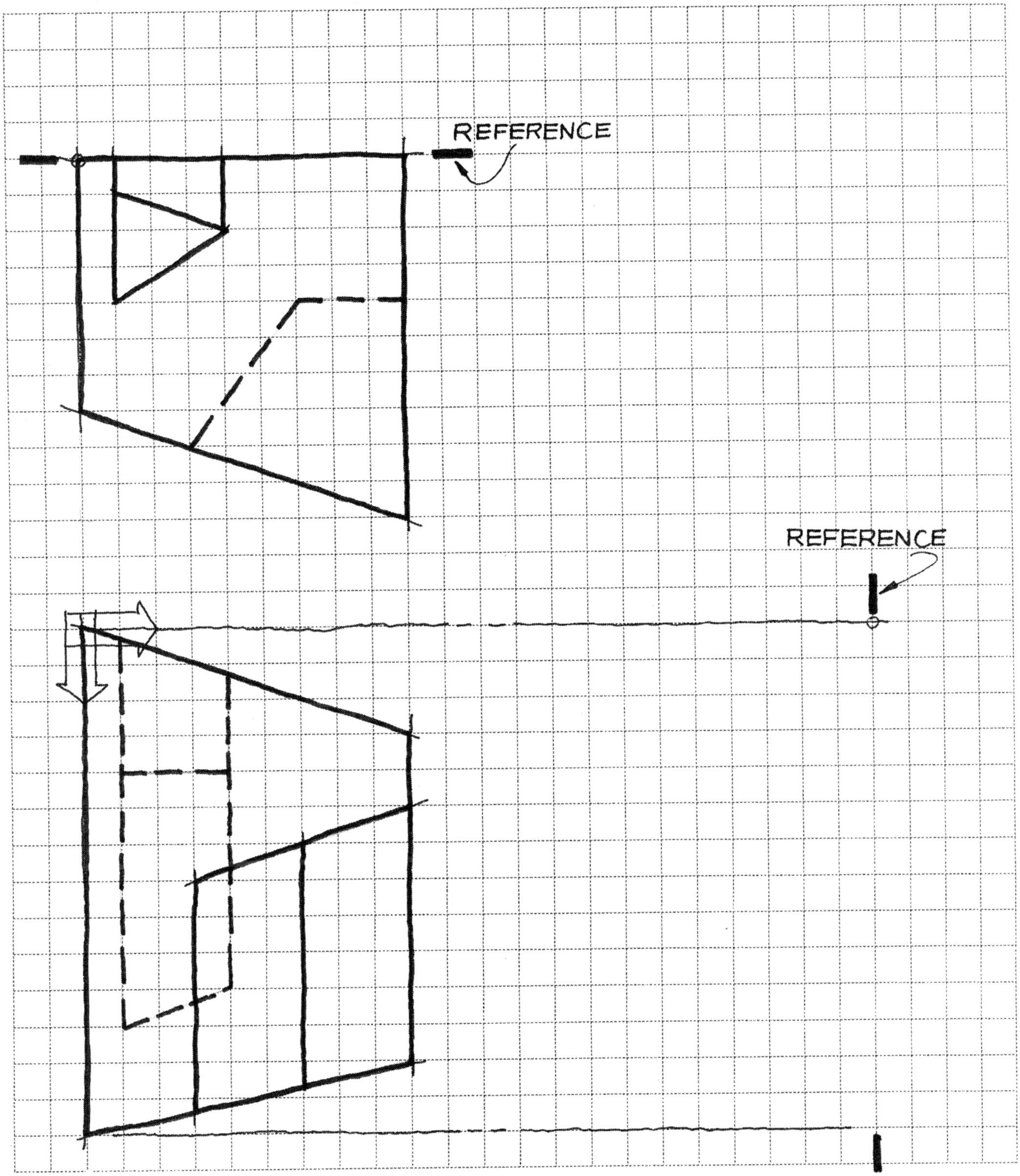

Drawn By:		FREEHAND SKETCHING	
	Scale	TITLE REFERENCE PLANES	
Date	Sheet of	Drawing Number A	6.14

COMPLETE THE FRONT VIEW OF THE ROOF AND CHIMNEY INTERSECTION.

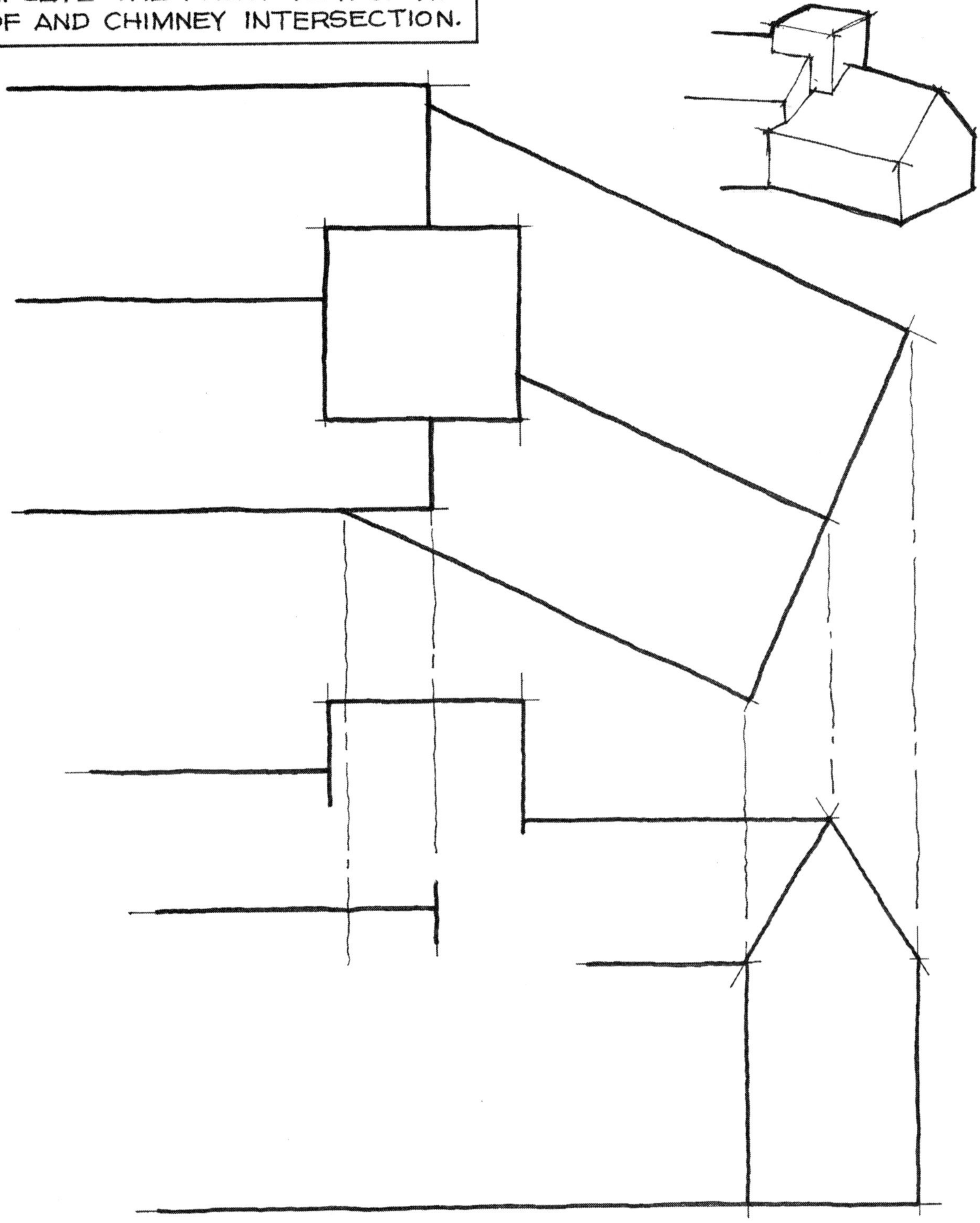

Drawn By:		FREEHAND SKETCHING	
	Scale	TITLE CUTTING PLANES	
Date	Sheet of	Drawing Number A	6.15

QUESTIONS:

1. Surface (C) in the top view appears as an edge in which view?
2. Surface (A) in the front view is shown as what surface in the top view?
3. What type of surface is (J)?
4. What material is the tool block made of?
5. How many inclined surfaces can be found on the tool block?
6. How many profile surfaces are on the block?
7. Surface (P) is what type of surface?
8. How many oblique surfaces can be found on the tool block?
9. Is surface (B) parallel to surface (O)?
10. Is surface (J) parallel to surface (G)?

ANSWERS:

1. ________
2. ________
3. ________
4. ________
5. ________
6. ________
7. ________
8. ________
9. ________
10. ________

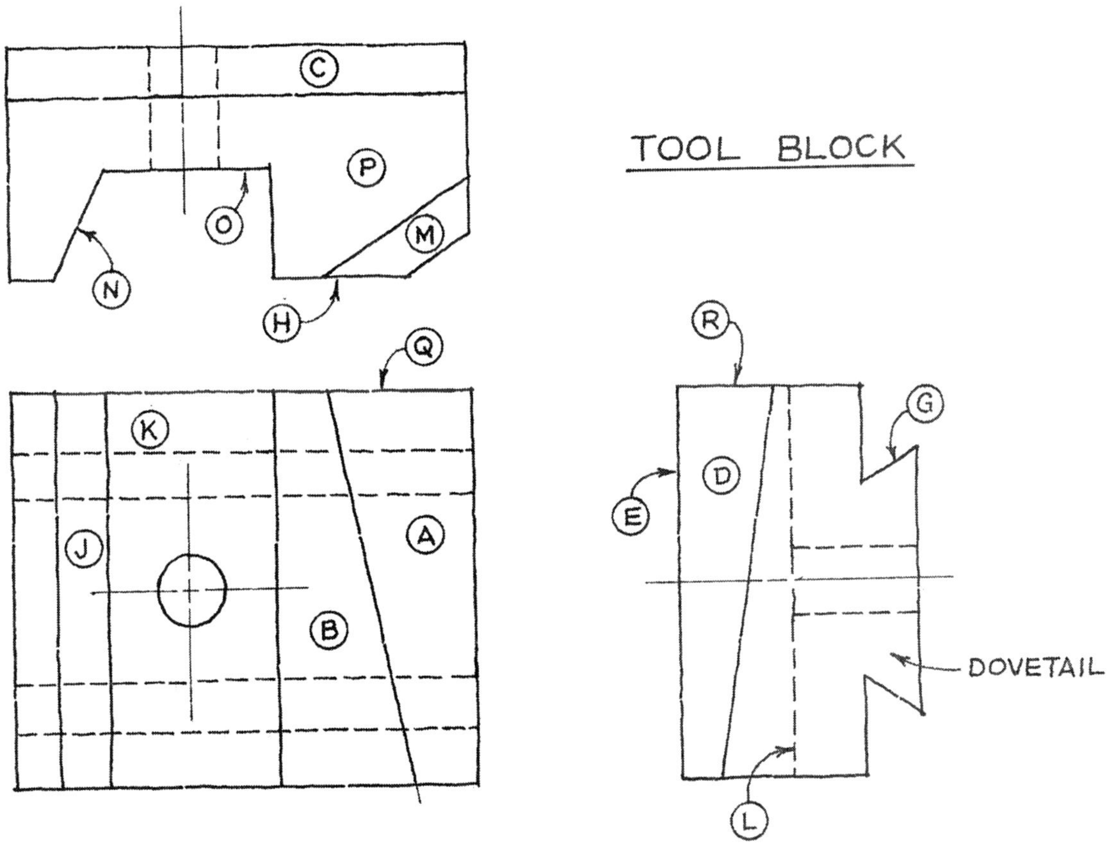

Drawn By:		FREEHAND SKETCHING	
	Scale	TITLE READING DRAWINGS	
Date	Sheet of	Drawing Number A	6.16

7 Auxiliary Views

Coordinated with VMGED Chapter 14

Auxiliary Views—General Conditions

An *auxiliary view* is a view taken at an angle not normal to an object's axis system. For example, the auxiliary view marked A1 in Fig. 7-1 is parallel to the object's depth (Z-axis) but is at an angle to UCS-Y and UCS-X. An auxiliary viewing position is taken to:

1. Facilitate the construction of geometry not perpendicular or parallel to the object's axis system.

2. Display angular or oblique geometry in normal position.

3. Display an object in pictorial (axonometric) position.

Note in Fig. 7-1 that view A1 establishes a local axis system normal (true shape) in its X and Y directions. Taking a correct viewing direction is the first and most important step in achieving an auxiliary view. **The view is determined by the viewer's direction of sight**. Auxiliary views are drawn so that measurements can be made, angles determined, and clearances found. Because much geometry in an auxiliary ends up being foreshortened, *partial auxiliary views*, like that shown in Fig. 7-2, are common. Sheet 7.0 (on page 145) shows the basic relationships in an auxiliary view. Fundamental is the use of a plane of reference to control spatial measurements perpendicular to the direction of sight. Study the placement of the plane on sheet 7.0 and its appearance relative to the direction of sight in central, and adjacent views. In all views, the plane of reference is parallel to the assumed direction of sight of the auxiliary view.

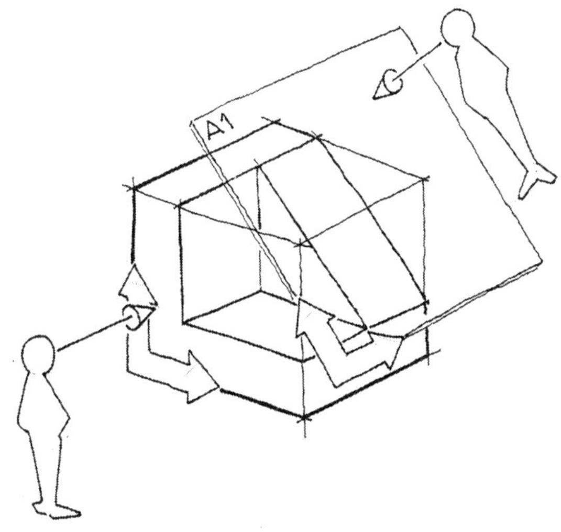

Fig. 7-1. The direction of sight determines the view.

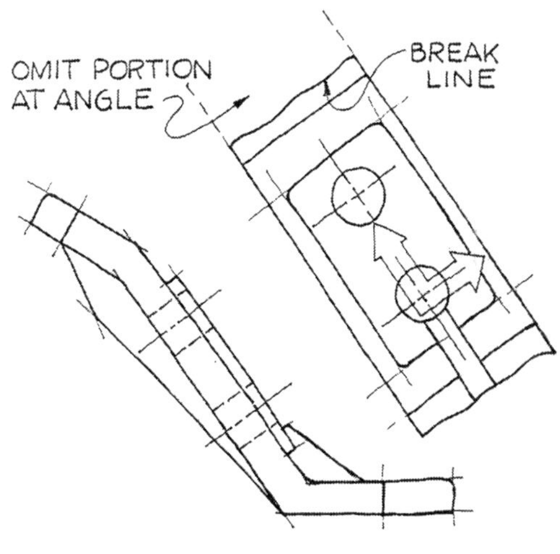

Fig. 7-2. Partial auxiliary view.

Primary and Secondary Auxiliary Views

A *primary auxiliary view* is one where the direction of sight is perpendicular to one of the object's axes and at an angle to the other two.

A *secondary auxiliary view* is one where the direction of sight is at an angle to all three principal object axes (Fig. 7-3).

Assignment: Complete Sheet 7.1. Use your main graphics text as a reference to determine definitions. In labeling the views, place labels in the order they are encountered spatially: nearest point to the farthest with the farthest point last.

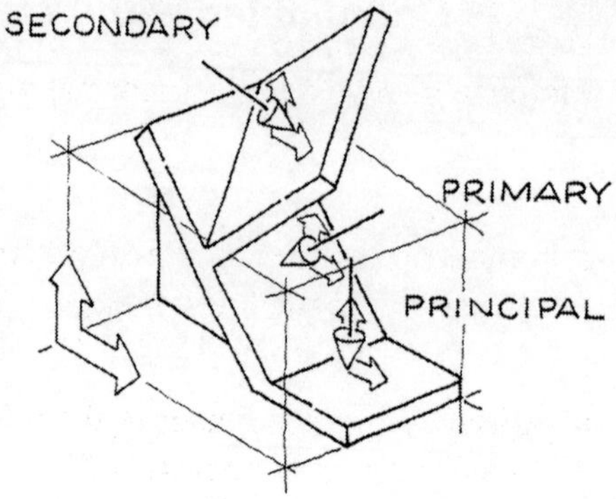

Fig. 7-3. Primary and secondary auxiliary views.

Understanding Auxiliaries

Lines (edges, limits, and intersections) parallel to a plane of reference remain parallel to that plane in any and all views. This allows auxiliary views to be easily drawn. Note in Fig. 7-4 that the top and bottom of the object are parallel to the reference plane as revealed in the front view. In the auxiliary view, the top and bottom (the height) remain parallel to the reference plane.

Not only are the edges parallel, but their distances perpendicular to the reference plane remain constant in all views. These distances are measurable in any view where the reference plane appears as an edge.

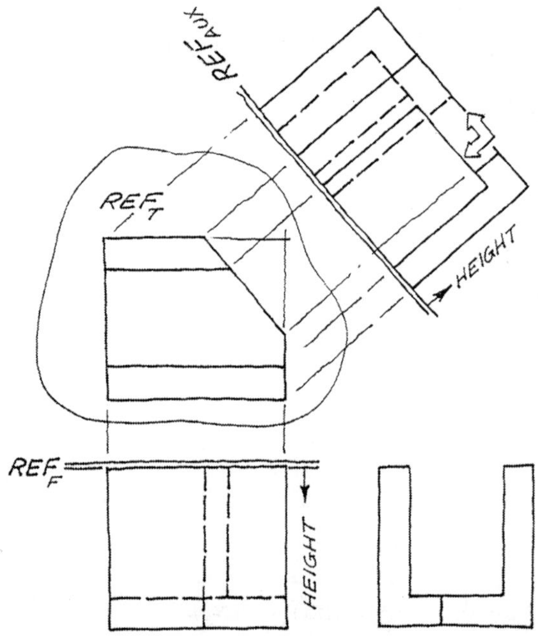

Fig. 7-4. Parallelism in auxiliary views.

Practical Example

Figure 7-5 presents the steps in creating an auxiliary view. In this example, a normal (true shape) view of the inclined surface is desired.

Step 1. Establish a viewing direction perpendicular to the inclined surface. This can be done in the side view where the inclined surface appears as an edge. The plane of reference that controls spatial measurements is parallel to the direction of sight. It appears normally in the side view and as an edge in the front and top views.

Step 2. Establish the edge views of the reference plane in the front and auxiliary views at convenient locations. Note the viewer looking straight down on the inclined plane in the side view and that plane turned 90 degrees in both the auxiliary and front views. You now have three successive views of the reference plane: edge view-normal view-edge view.

Step 3. Project points and vertices into the auxiliary view from the side view, parallel to the lines of sight.

Step 4. Measure distances perpendicular to the reference in the front view. Lay these measurements off along their corresponding projectors in the auxiliary view, always measuring from the reference plane in the same direction (in this case, toward the central view).

Step 5. Assemble the object in the auxiliary view from near to far, determining visibility as you go along.

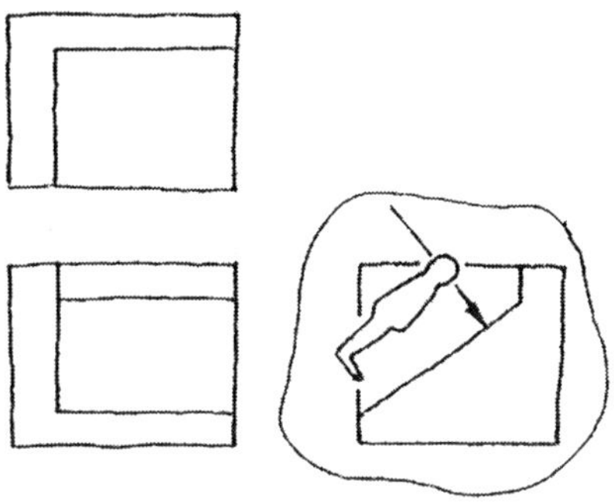

Step 1

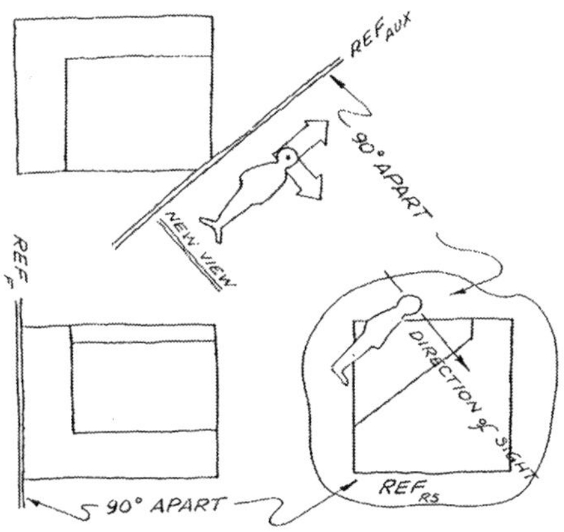

Step 2

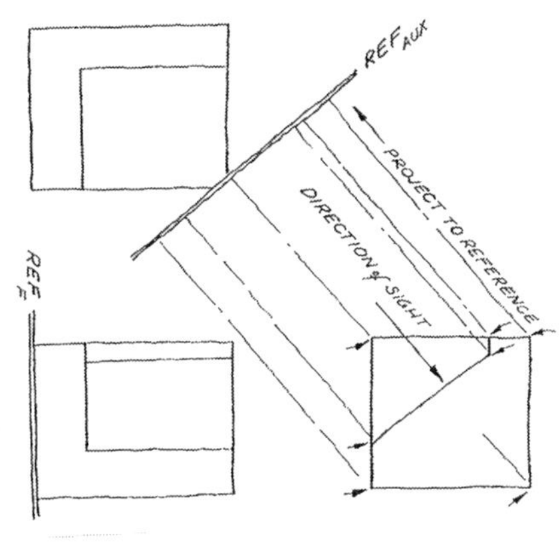

Step 3

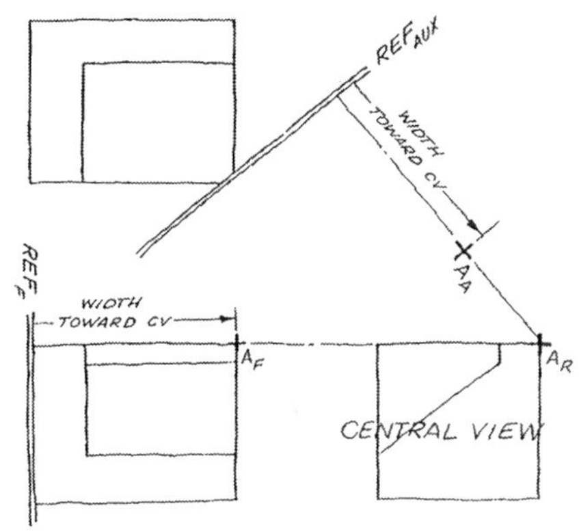

Step 4

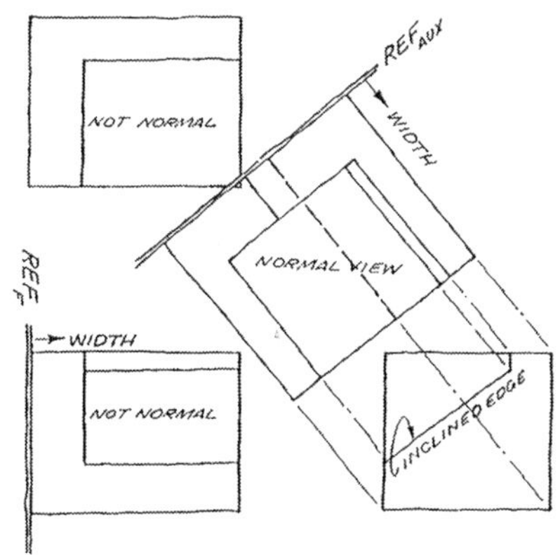

Step 5

Fig. 7-5. Steps in completing an auxiliary view.

CAD Example

The same techniques just discussed to sketch auxiliaries are used in 2D CAD with the addition of CAD's unique tools. Parallel lines can be created through points or at a user-specified perpendicular distance from a base (reference) line. Distances are easily measured and transferred.

Modeling in 3D CAD offers the opportunity to view the geometry from any direction. Viewing directions are usually defined by a vector normal to a surface. In Fig. 7-6, an oblique surface defines a new view. Note that the UCS is placed on the oblique plane, ensuring constructions will be normal to that surface. Figure 7-7 has aligned the view to the UCS so the surface appears normally.

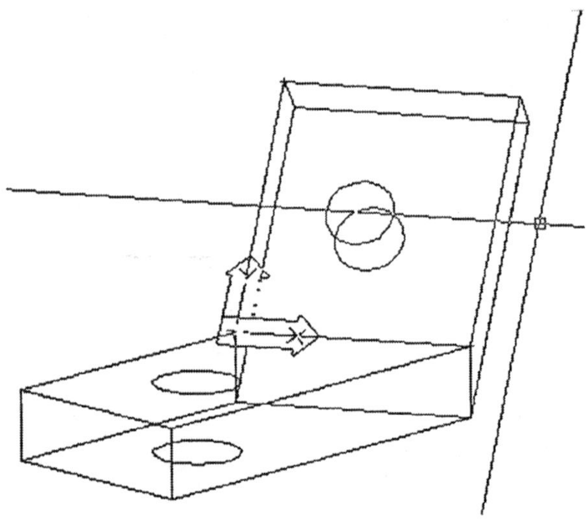

Fig. 7-6. Position of UCS normal to oblique plane.

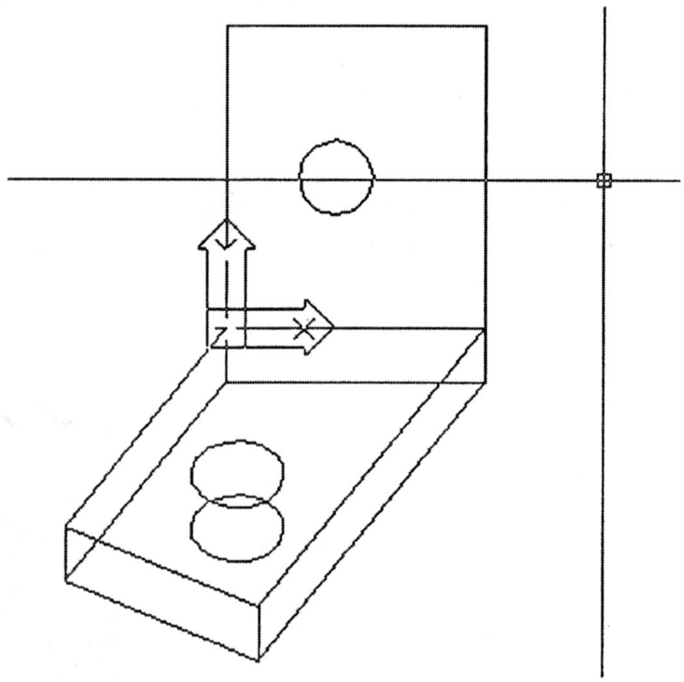

Fig. 7-7. View in Z direction of the local UCS.

AUXILIARY VIEW

REFERENCE PLANE METHOD

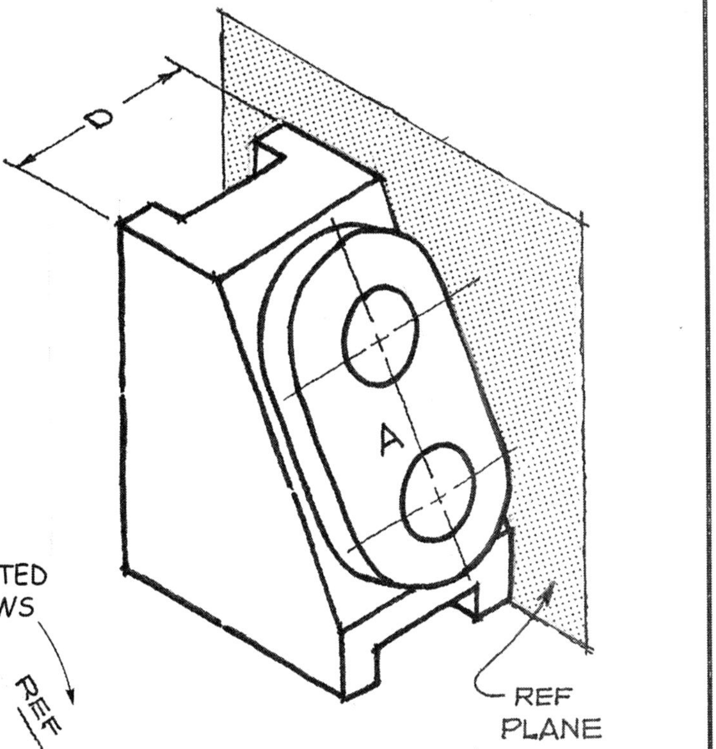

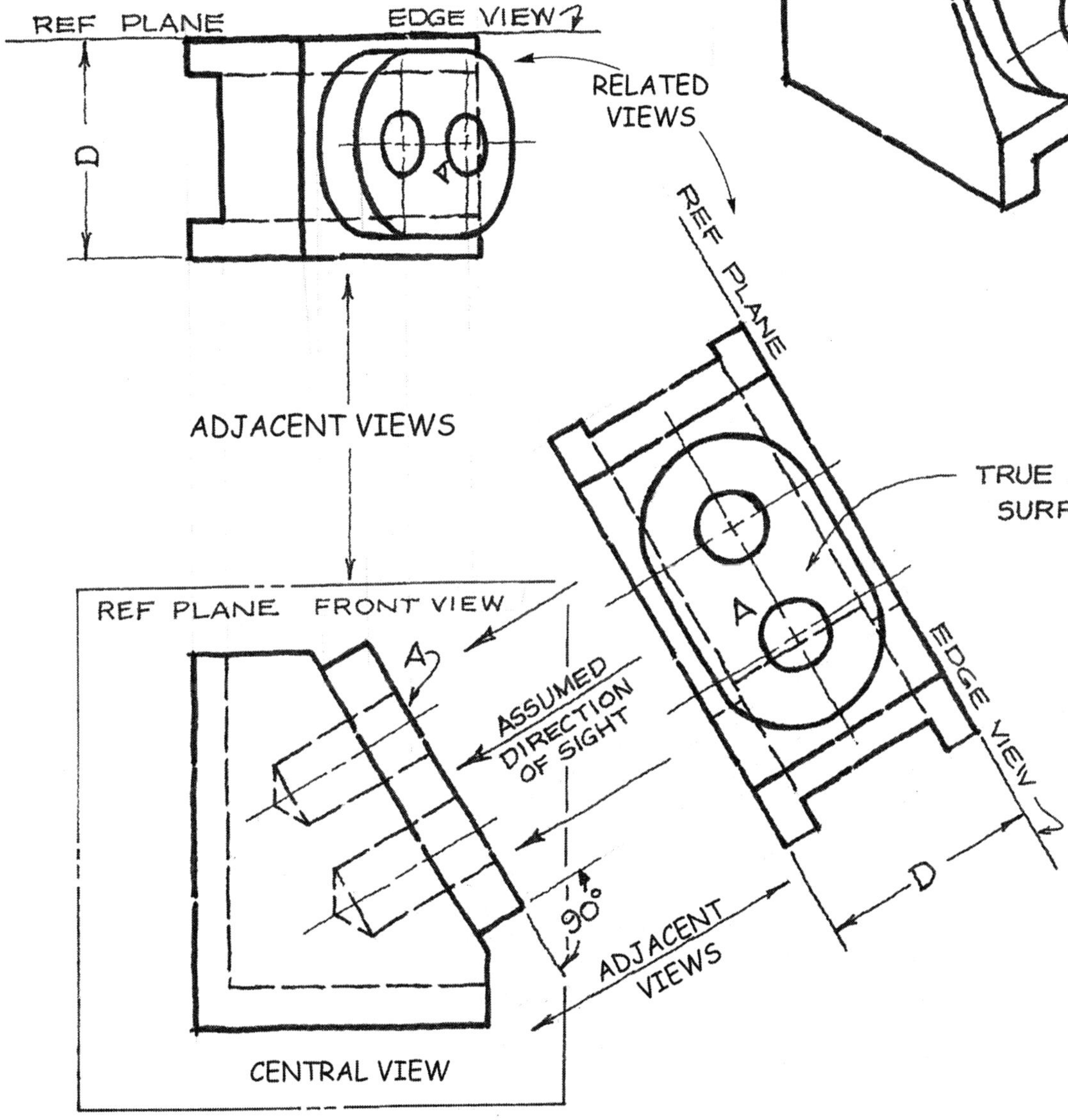

Drawn By:		FREEHAND SKETCHING	
	Scale	TITLE SAMPLE	
Date	Sheet of	Drawing Number A	7.0

A. Define the following terms.

line ______________________________

point ______________________________

plane ______________________________

normal view ______________________________

point view ______________________________

edge view ______________________________

reference plane ______________________________

cutting plane ______________________________

principal view ______________________________

auxiliary view ______________________________

adjacent view ______________________________

related view ______________________________

B. Fully label and subscript the following views.

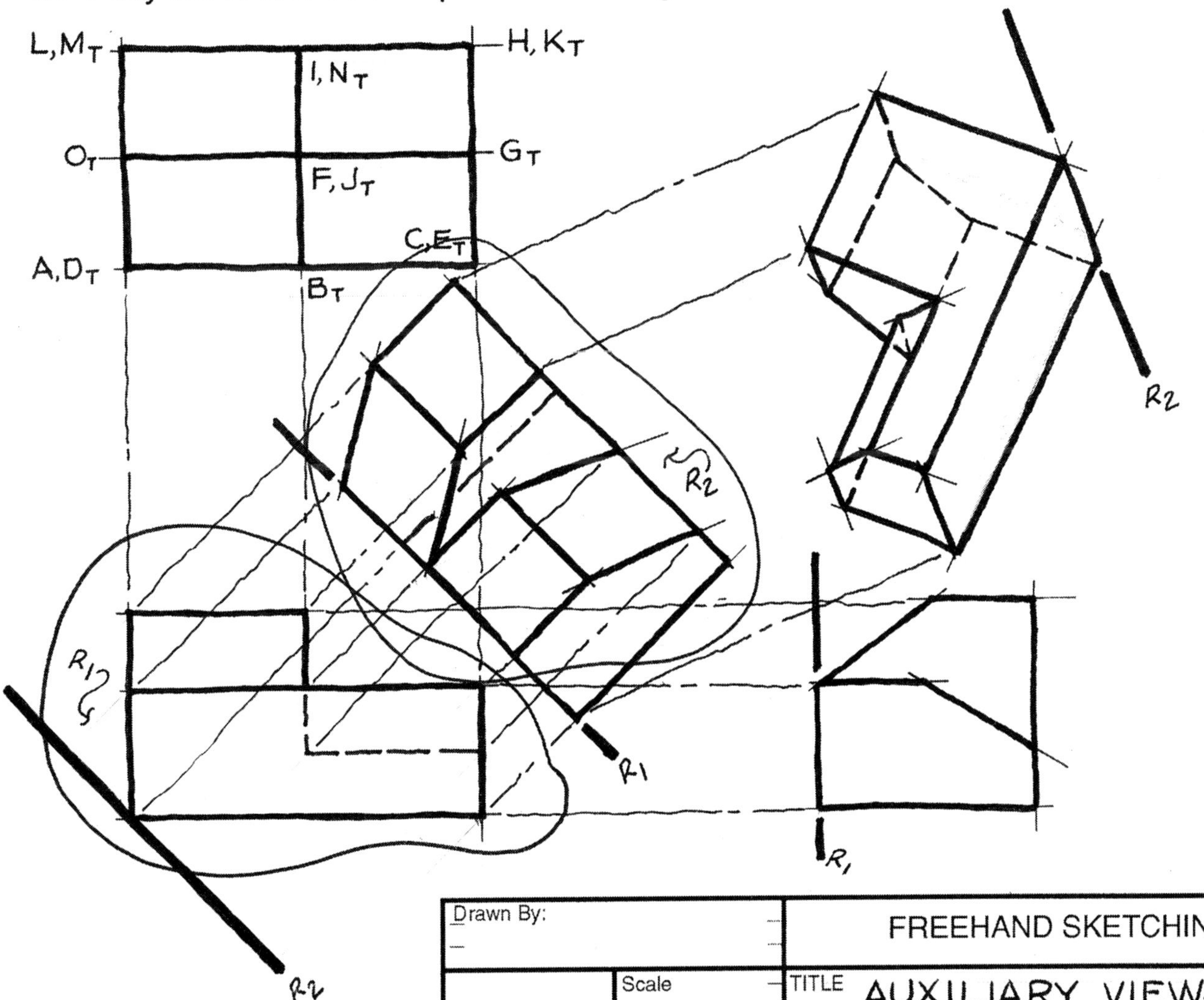

Drawn By:		FREEHAND SKETCHING	
	Scale	TITLE AUXILIARY VIEWS	
Date	Sheet of	Drawing Number A	7.1

CONSTRUCT A PARTIAL OR A COMPLETE AUXILIARY VIEW OF THE OBJECTS BELOW, AS SPECIFIED.

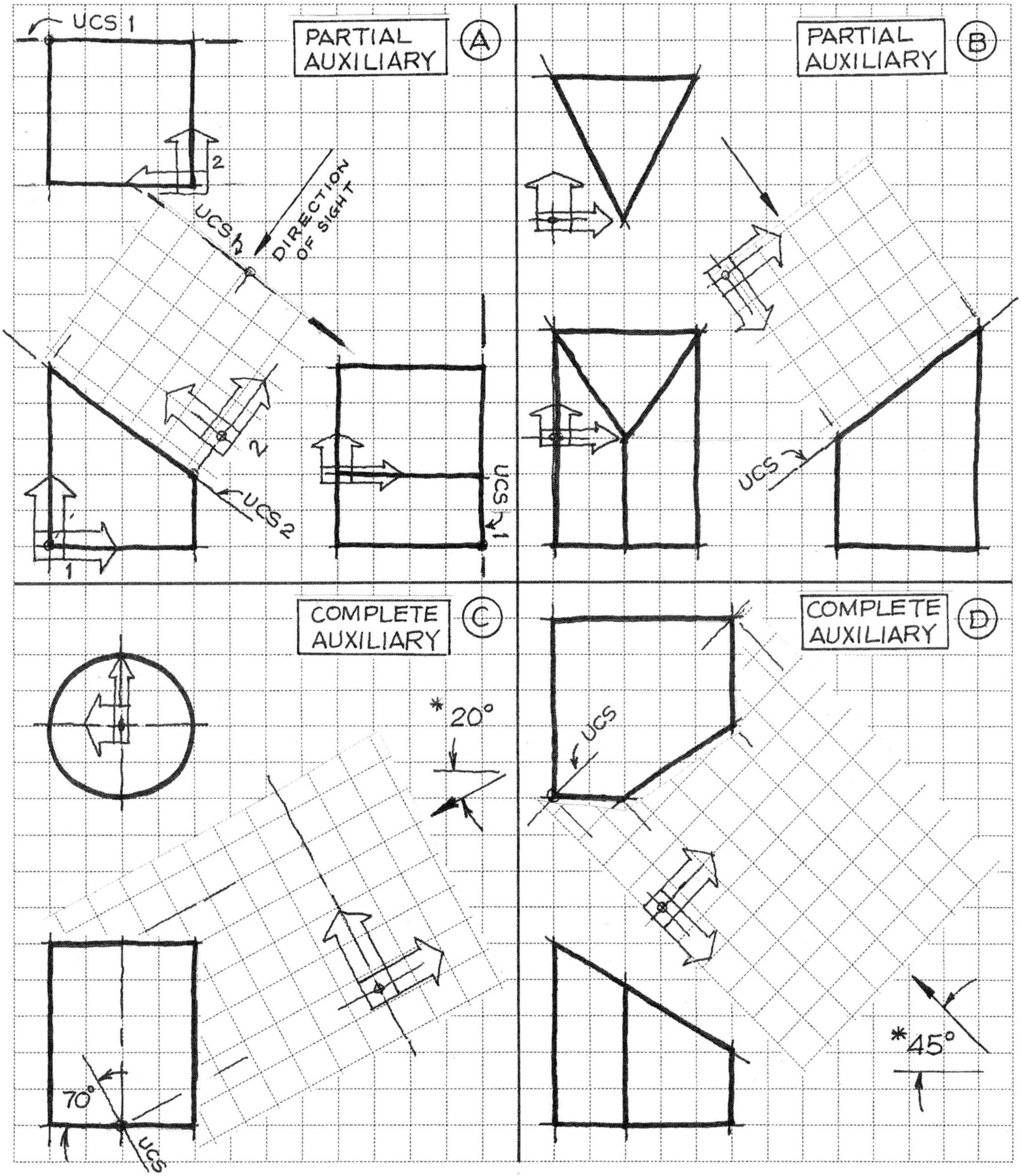

* UCS ROTATION

Drawn By:		FREEHAND SKETCHING	
	Scale	TITLE AUXILIARY VIEWS	
Date	Sheet of	Drawing Number A	7.2

COMPLETE MISSING VIEWS AND AUXILIARY VIEWS OF INCLINED SURFACES. ATTACH UCS AS REQUIRED.

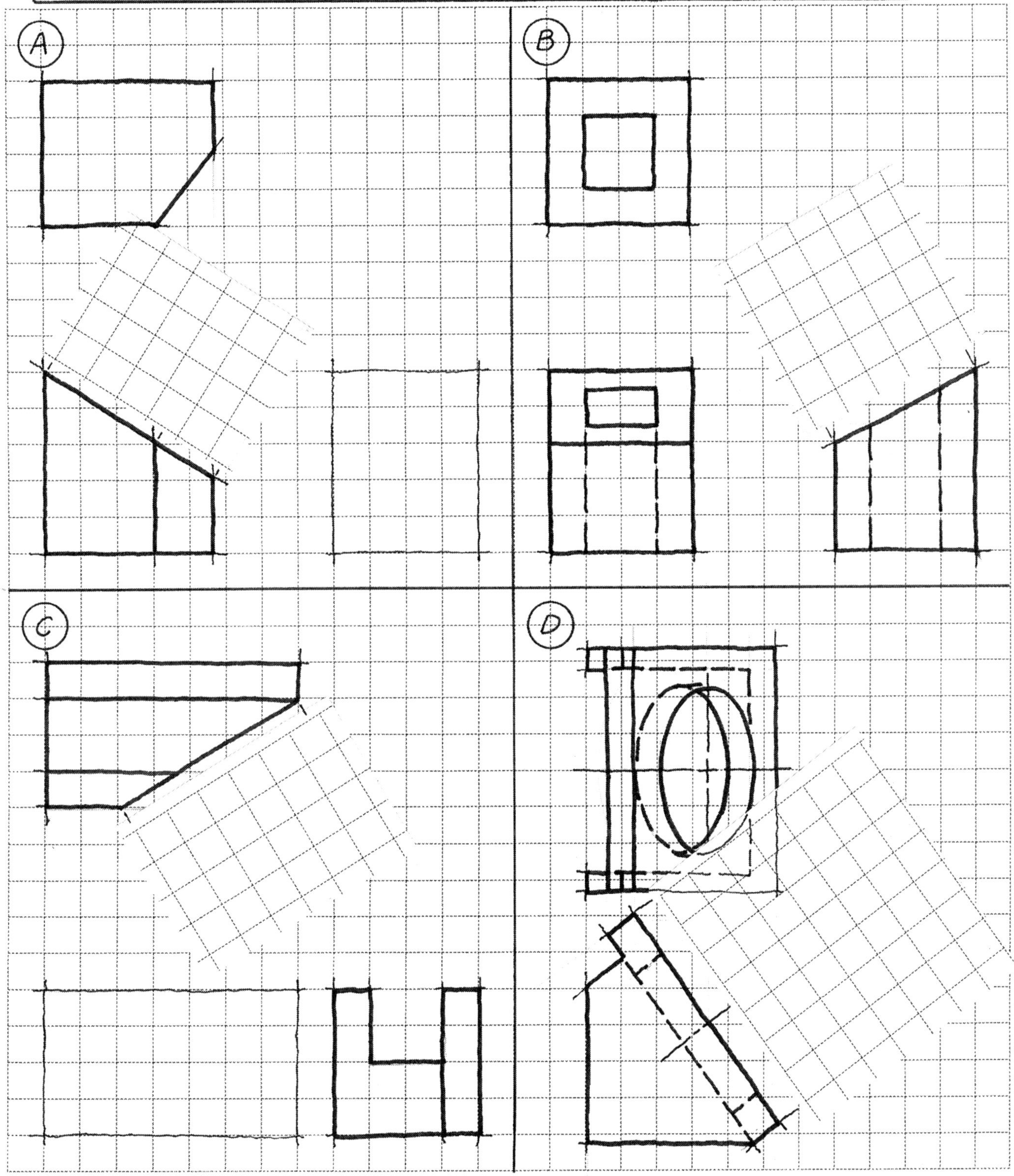

Drawn By:		FREEHAND SKETCHING	
	Scale	TITLE AUXILIARY VIEWS	
Date	Sheet of	Drawing Number A	7.2.1

COMPLETE A PRIMARY AUXILIARY VIEW OF THE ENTIRE OBJECT THAT INCLUDES A NORMAL VIEW OF THE TOP SURFACE.

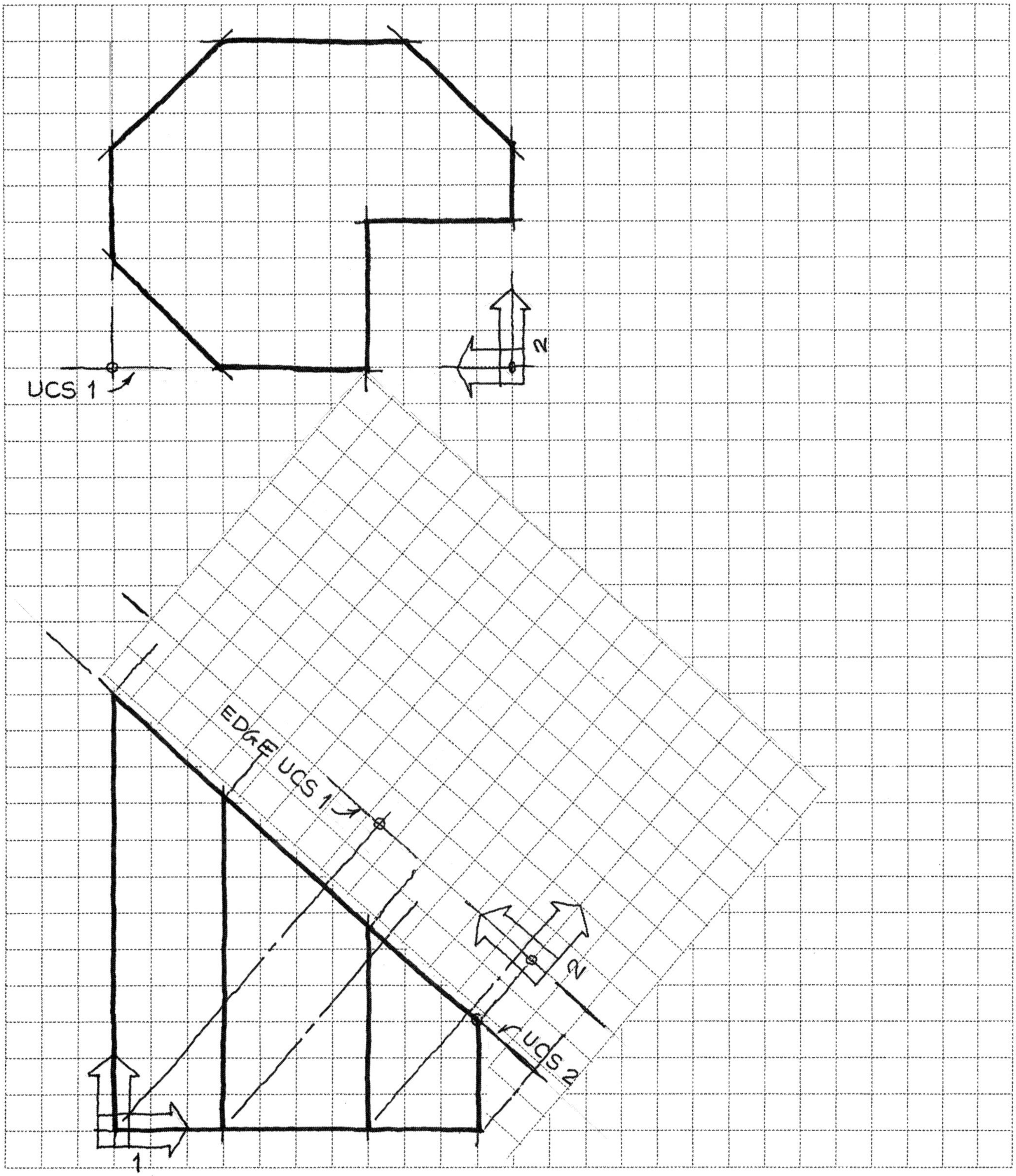

Drawn By:		FREEHAND SKETCHING	
	Scale	TITLE AUXILIARY VIEWS	
Date	Sheet of	Drawing Number A	7.3

COMPLETE A PRIMARY AUXILIARY VIEW OF THE ENTIRE OBJECT THAT INCLUDES A NORMAL VIEW OF THE TOP SURFACE.

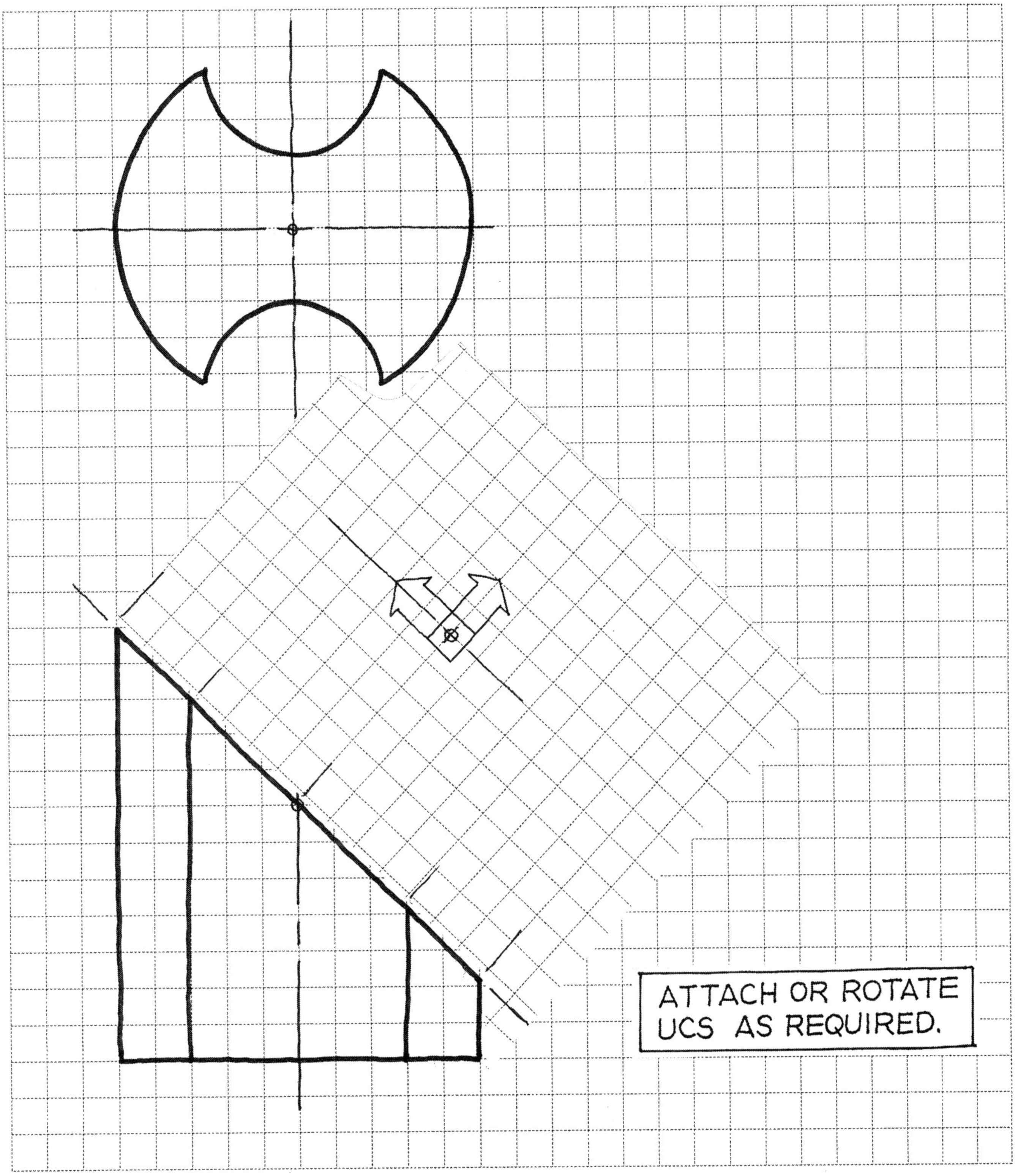

ATTACH OR ROTATE UCS AS REQUIRED.

Drawn By:		FREEHAND SKETCHING	
	Scale	TITLE AUXILIARY VIEWS	
Date	Sheet of	Drawing Number A	7.4

CONSTRUCT TWO COMPLETE AUXILIARY VIEWS OF THE OBJECT IN THE AREAS SHOWN. REFER TO SHEET 7.1.

ROTATE OR ATTACH UCS AS REQUIRED

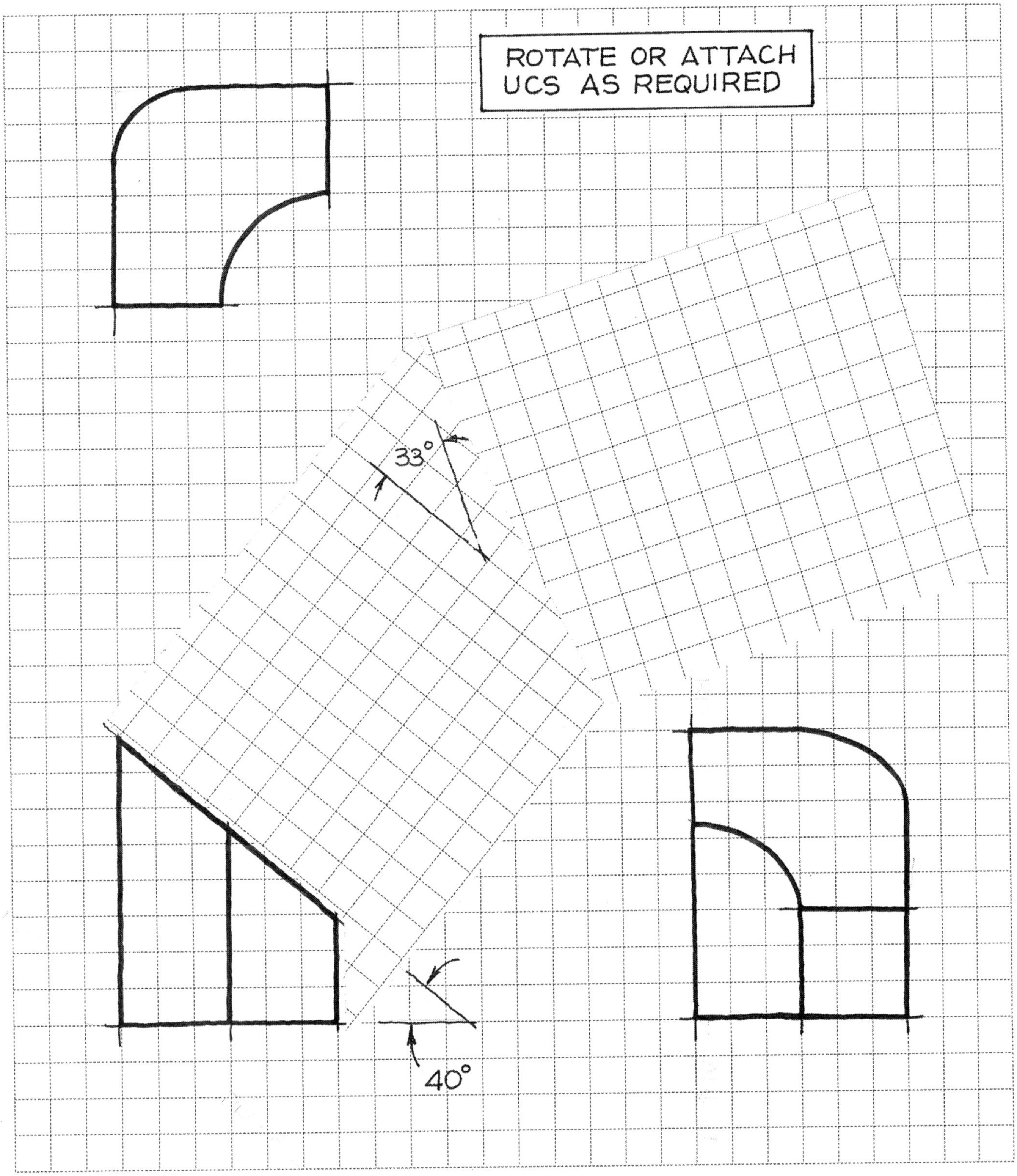

Drawn By:		FREEHAND SKETCHING	
	Scale	TITLE AUXILIARY VIEWS	
Date	Sheet of	Drawing Number A	7.5

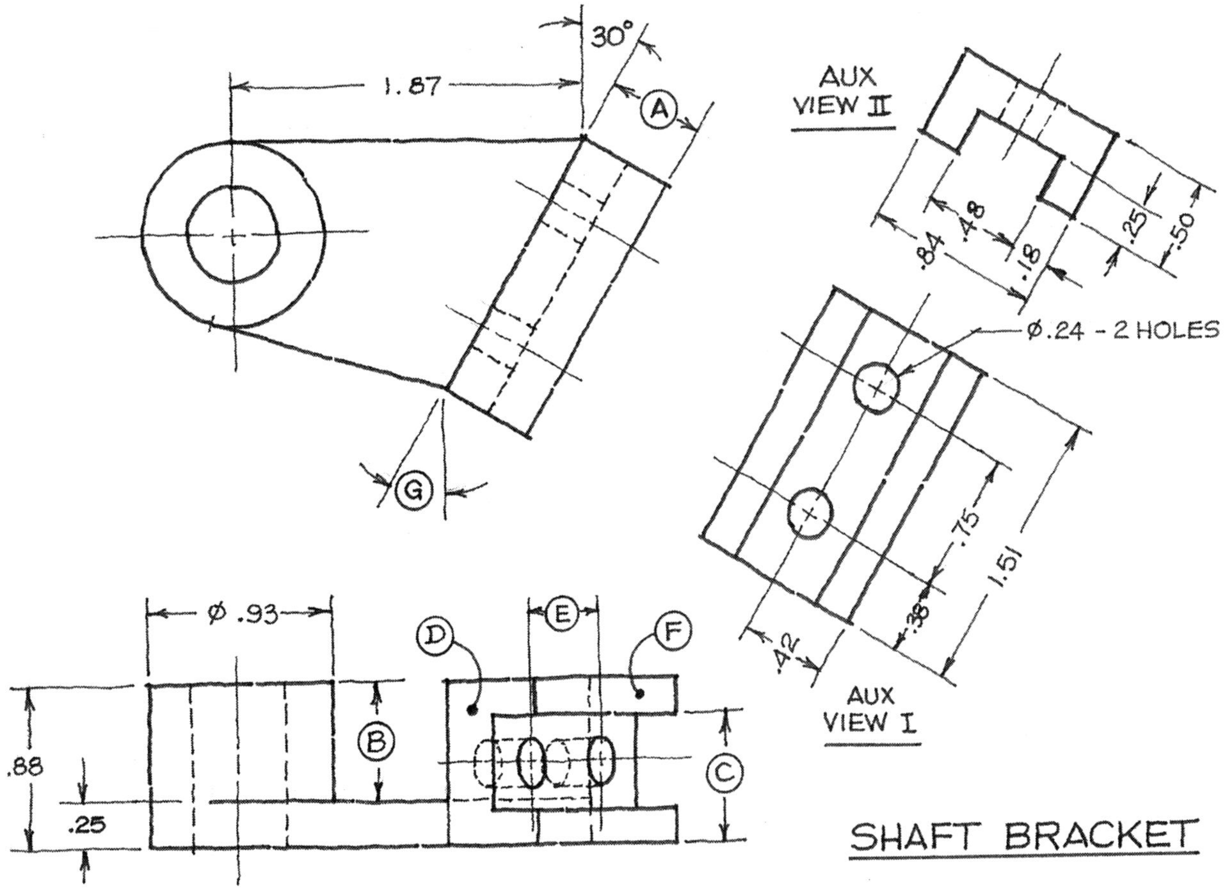

SHAFT BRACKET

QUESTIONS:

ANSWERS:

1. What is the diameter of the cylindrical end of the shaft? — 1. __________
2. What is dimension (A)? — 2. __________
3. What type of surface is (D) in the front view? — 3. __________
4. What is dimension (B)? — 4. __________
5. The actual distance between holes shown at (E) is _____ and can be accurately measured in which view? — 5. __________
6. Surface (F) can be seen in true shape in which view? — 6. __________
7. The depth of the slot is ____ and can be found in which view? — 7. __________
8. What is angle (G)? — 8. __________

Drawn By:		FREEHAND SKETCHING	
	Scale	TITLE READING DRAWINGS	
Date	Sheet of	Drawing Number A	7.6

8 Sectional Views

Coordinated with VMGED Chapter 13

Sectional Views—General Conditions

A *sectional view* is produced by passing a cutting plane through an object, taking a viewing direction normal to that plane, and removing the part of the object between the viewer and the cut. The result is a view of the object's interior detail. Objects with interior detail often require sectioning in order to allow easy visualization of cross-sectional profiles. Material in contact with the cutting plane is designated by a pattern of parallel lines called crosshatching. Objects without interior detail do not require sectioning.

Study sheet 8.0 in detail (page 167). The location of the section is generally shown by the cutting plane-on-edge. The direction of the sectional view is shown by direction of sight arrows perpendicular to the cutting plane.

The cutting plane should be as straight as possible, without excessive bends and offsets. It should never double back on itself. The standard representation of the plane is a bold, dark line, often with two dashes.

When a section appears in normal aligned position (as with a front view in section) and is the same scale as the other views, and the location of the cutting plane is unambiguous, the cutting plane may be omitted and the section not named (Fig. 8-1). However, if the location of the section is unclear, if the section has been removed from its expected position, or if the section's scale has been changed, the cutting plane should be identified and the sectional view named.

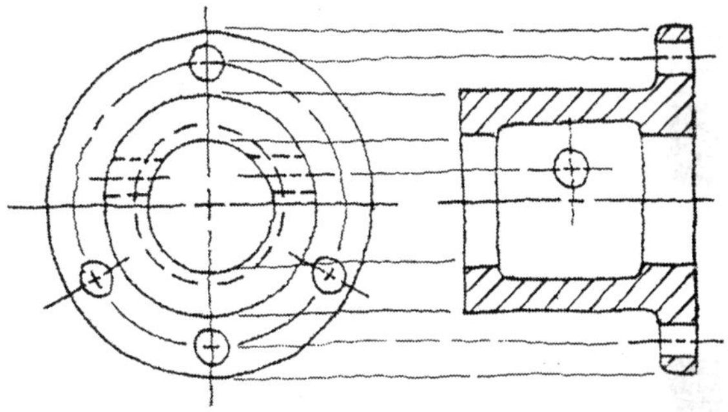

Fig. 8-1. Sectional view in principal position.

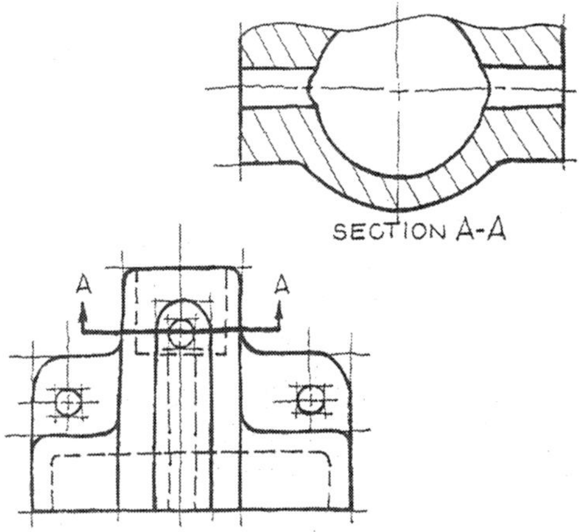

Fig. 8-2. Sectional view in removed position with a change of scale.

Assignment: Find a small object that has interior detail and can fit in the palm of your hand. Complete orthographic and pictorial sketches of the object in section. Use basic proportions.

Types of Sectional Views

When a cutting plane is passed completely through an object (Fig. 8-3) the view is called a *full section*. Full sections are appropriate for objects whose internal features are nonsymmetrical and essentially extend from side-to-side.

A *half-section* is appropriate to show symmetrical detail (Fig. 8-4). A half-section has the added advantage of being able to show the exterior detail in the same view as the section. No sharp edge is shown at the centerline.

A *revolved section* (Fig. 8-5) is used to show the cross-sectional shape of a rib, spoke, or other connecting feature. The section is shown on the object, in the position where it has been taken, so no cutting plane designation is necessary.

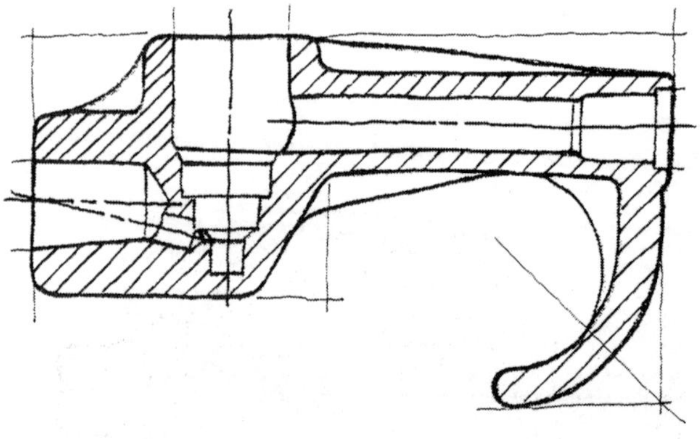

Fig. 8-3. A full section.

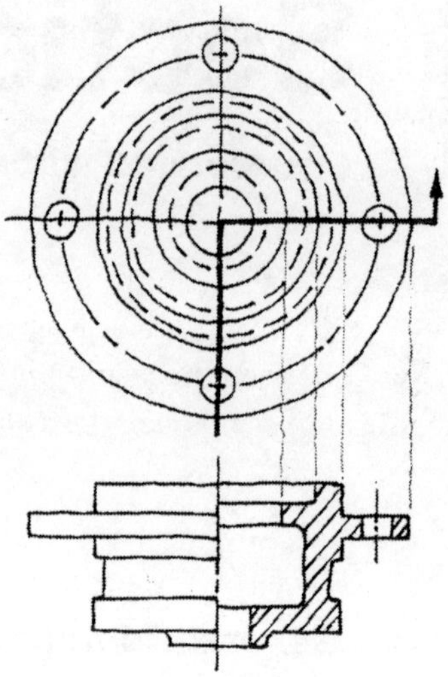

Fig. 8-4. A half section.

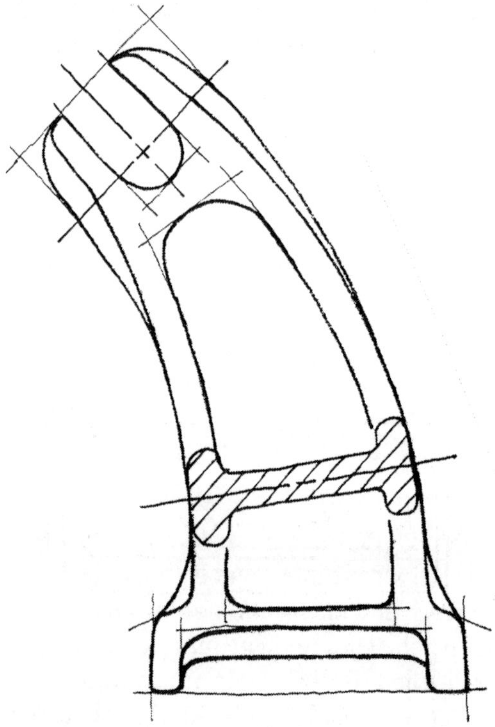

Fig. 8-5. A revolved section.

A *removed section* (Fig. 8-6) functions like a revolved section for objects whose profiles might be highly variable. The section is revolved but then moved off the view. This may necessitate the use of a cutting plane and section name.

An *offset section* is used when features do not lie in the same plane (Fig. 8-7). Sharp edges caused by bends in the cutting plane are ignored when the section itself is drawn. The cutting plane must not bend back on itself.

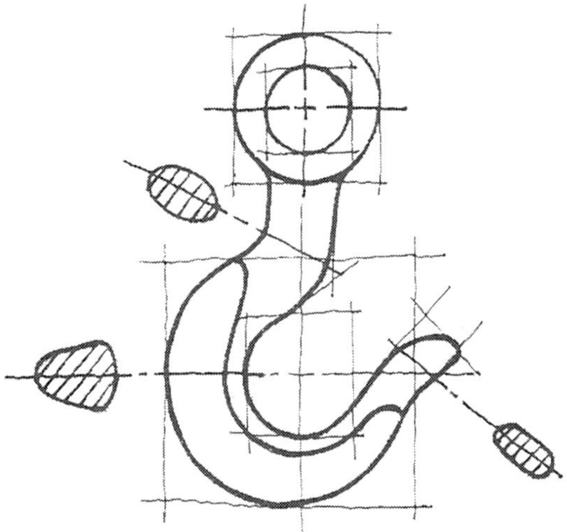

Fig. 8-6. Removed sections.

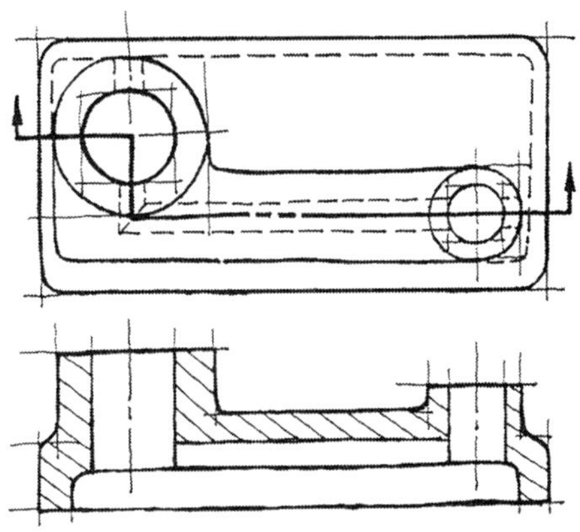

Fig. 8-7. An offset section.

A *broken section* (Fig. 8-8) is used to show interior detail at a specific location without the entire object being sectioned. The cutting plane is not shown and the section is not named.

A *sectional assembly* shows an assembly by representing parts together (Fig. 8-9). Note the different crosshatching used for various parts; parts with no interior detail are not sectioned.

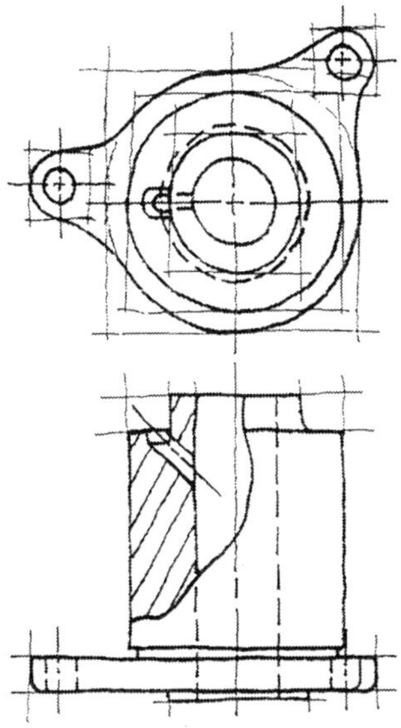

Fig. 8-8. A broken section.

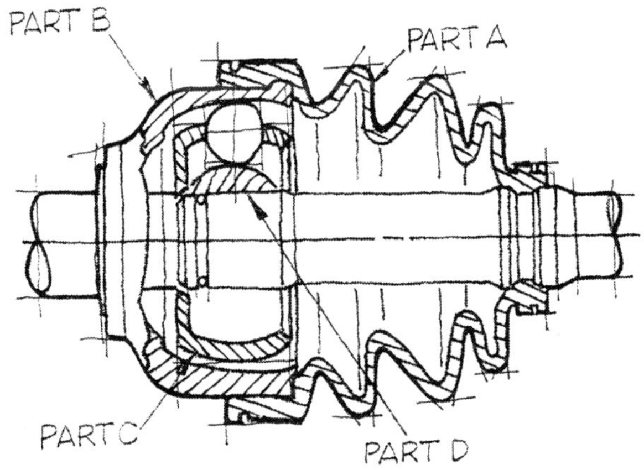

Fig. 8-9. A sectional assembly.

The Basic Rules of Sectioning

1. Do not section anything without interior detail. This includes rods, shafts, spheres, and other solid forms. See Fig. 8-10(a). Use a broken section for small details in otherwise solid objects.

2. Crosshatch only material that touches the cutting plane. See Fig. 8-10(b).

3. Revolve all radial features into the plane of the section. This includes ribs and holes as shown in Fig. 8-10(c).

4. Section features that are "full-body depth" or are tangent to sections that are full-body depth. Do not section small appendages, as shown in Fig. 8-10(d).

5. Do not section ribs, spokes, or other connections, as shown in Fig. 8-10(e).

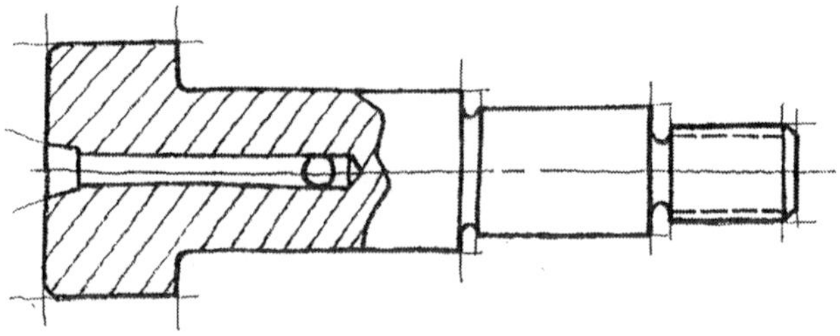

Fig. 8-10(a). A basic rule of sectioning—interior detail.

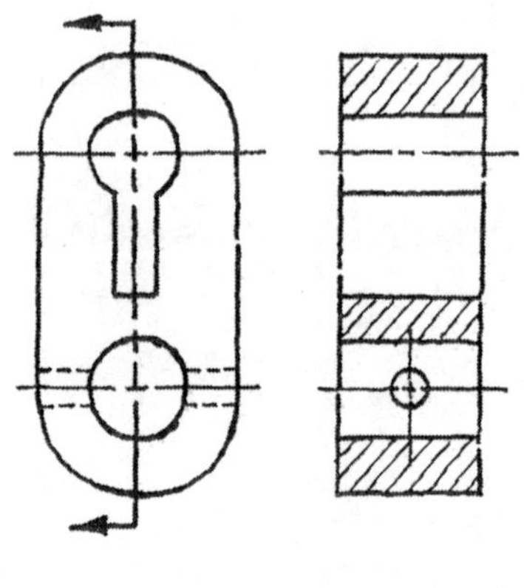

Fig. 8-10(b). A basic rule of sectioning—crosshatching.

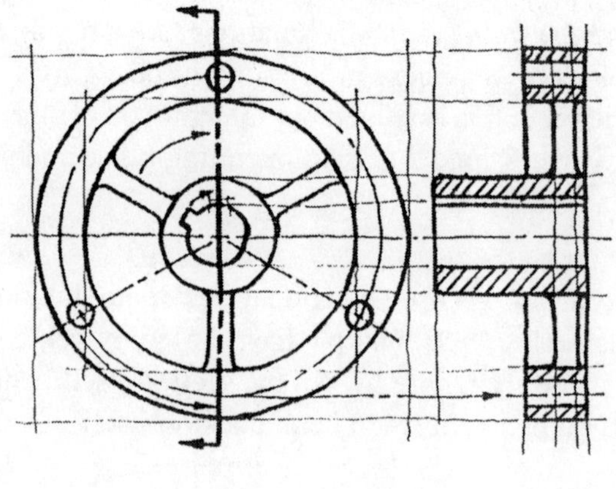

Fig. 8-10(c). A basic rule of sectioning—revolution of radial features.

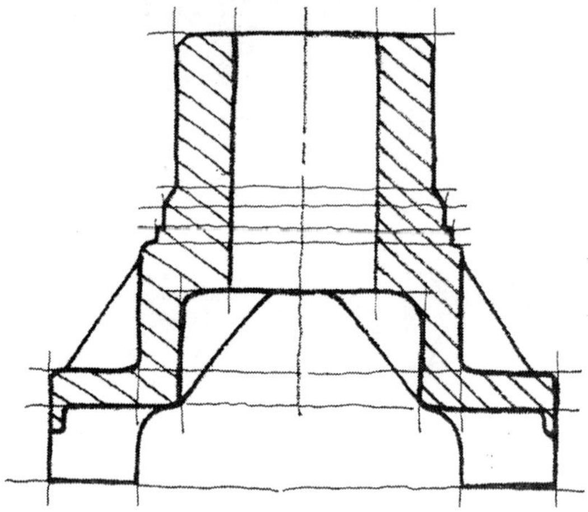

Fig. 8-10(d). A basic rule of sectioning—sectioning features not full-body depth.

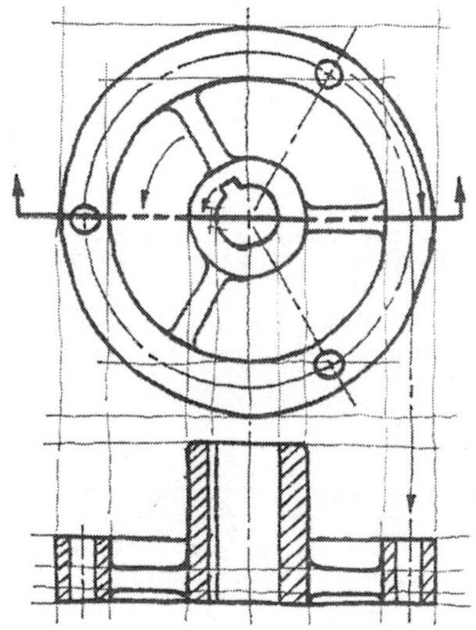

Fig. 8-10(e). A basic rule of sectioning—ribs or spokes.

6. Different parts must be sectioned differently. Section lines for an object are drawn the same direction anywhere the object appears on a drawing, as shown in Fig. 8-10(f).

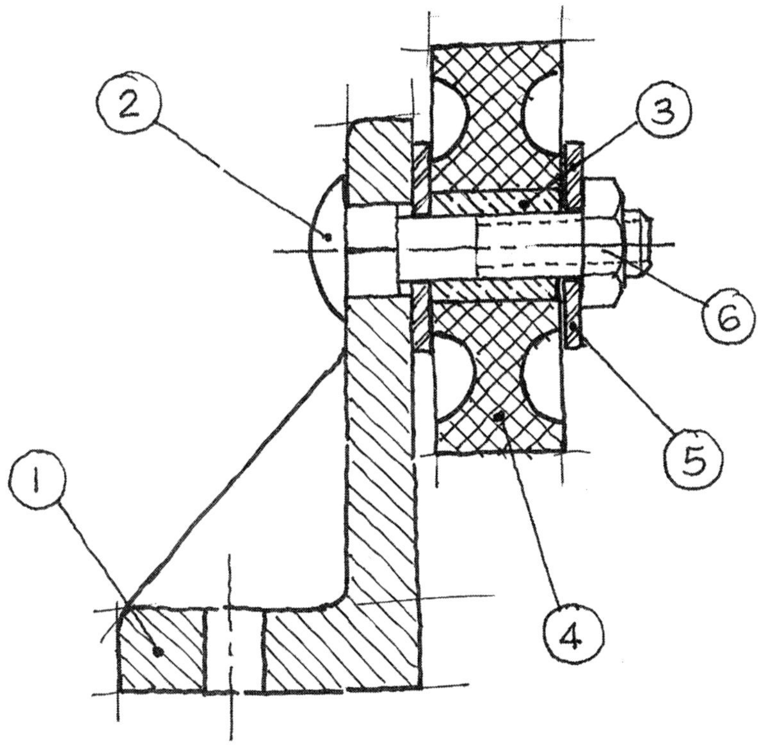

1	BASE
2	C-BOLT #
3	BUSHING #
4	ROLLER
5	WASHER (2)
6	HEX-NUT

Fig. 8-10(f). A basic rule of sectioning—section line pattern.

CAD Example—2D

Two-dimensional CAD uses polygon fills to represent sections. Figure 8-11 shows a front view before and after sectioning operations. Note that dashed lines are changed to solid lines. These solid boundaries are then filled with a crosshatch pattern.

CAD Example—3D

Three-dimensional CAD handles sections in different ways. In its most simple form, a wire frame model is cut by a plane (Fig. 8-12), resulting in wires that have to be joined and hatched. Solid models (Fig 8-13) can be directly sectioned, resulting in a more realistic appearance. However, note in the example that the convention of revolving radial features into the plane of the section has not been automatically followed, a limitation of CAD sectioning.

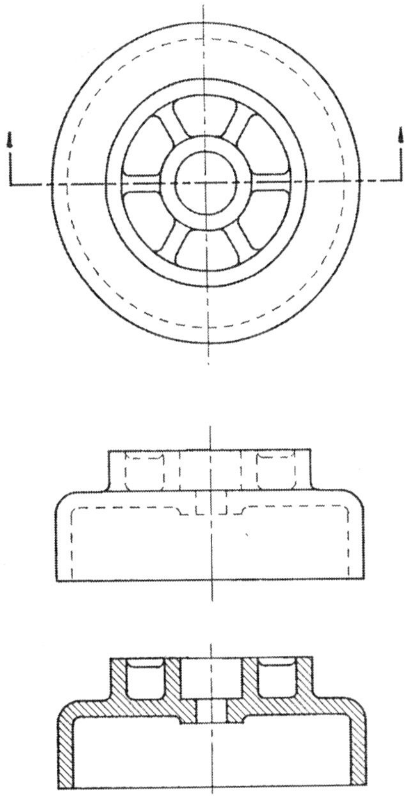

Fig. 8-11. 2D CAD sectioning.

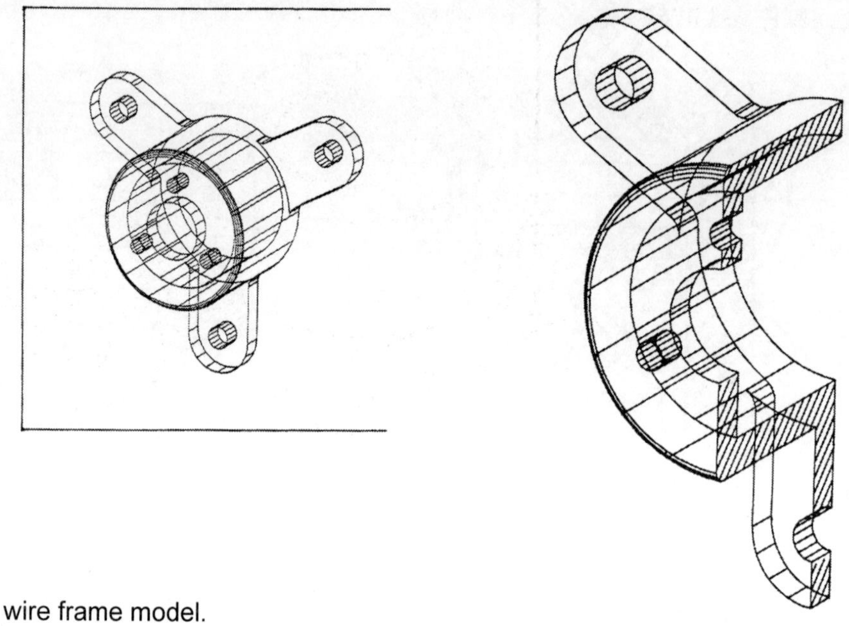

Fig. 8-12. Sectioned wire frame model.

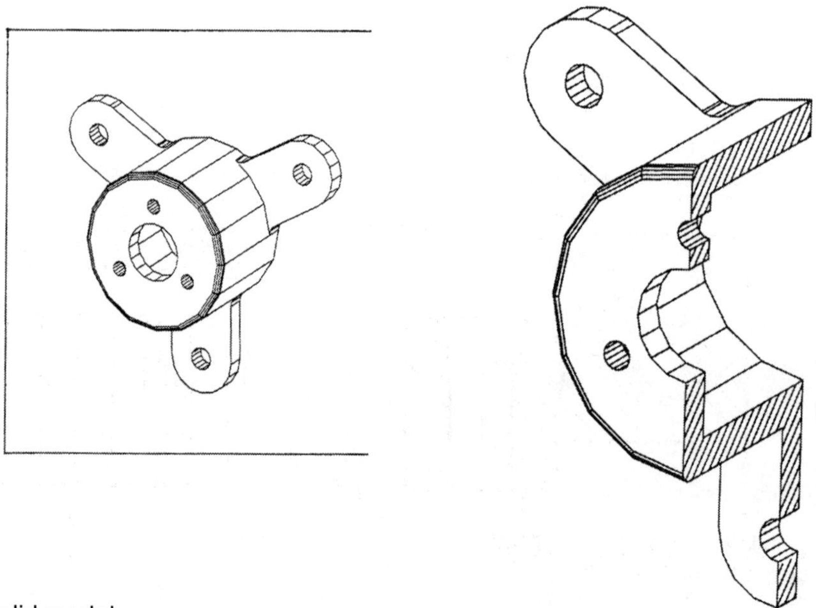

Fig. 8-13. Sectioned solid model.

SECTIONAL VIEW

CUTTING PLANE - UCS
PRINCIPAL

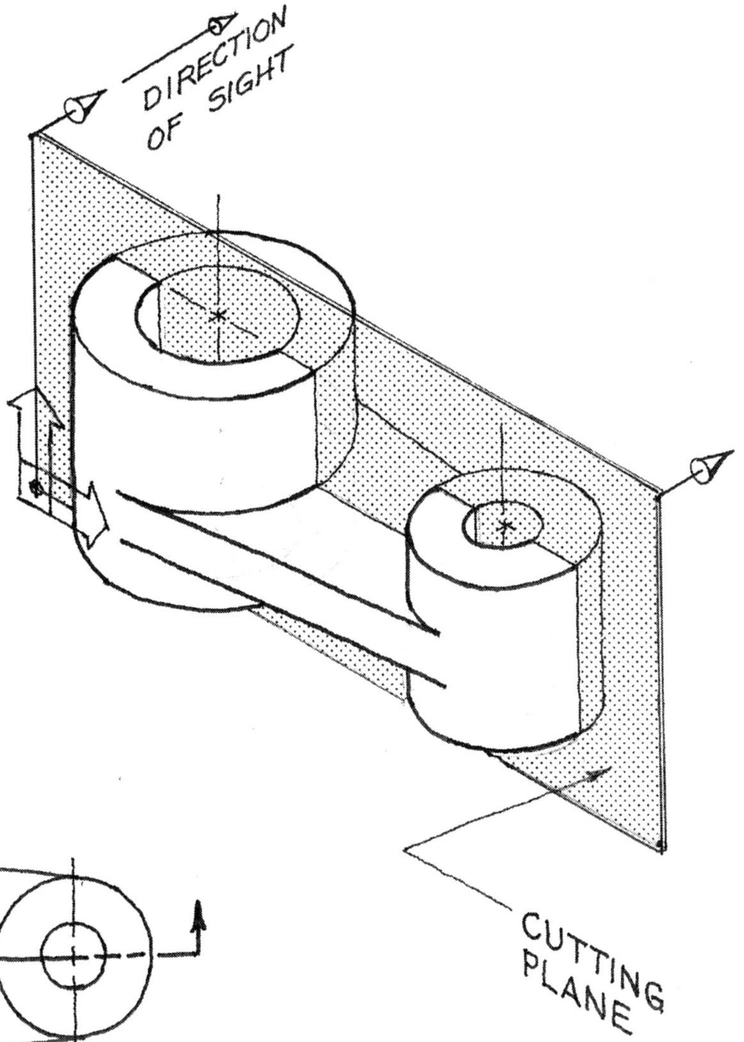

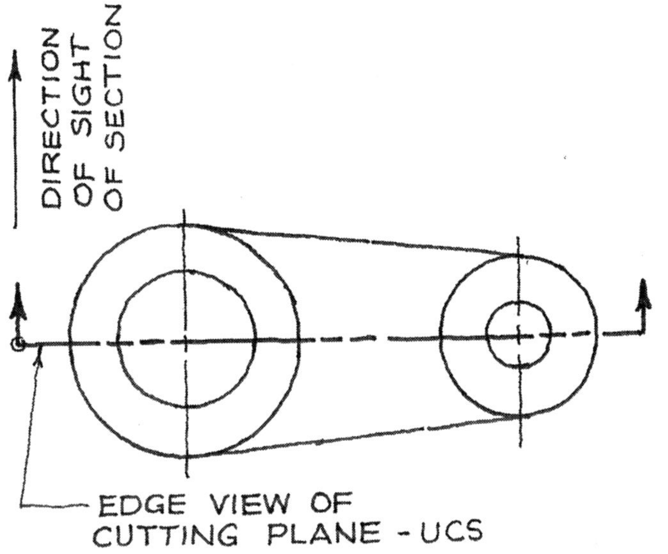

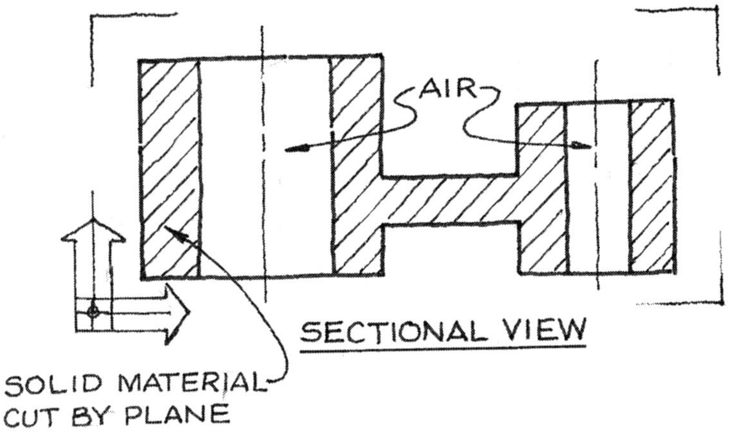

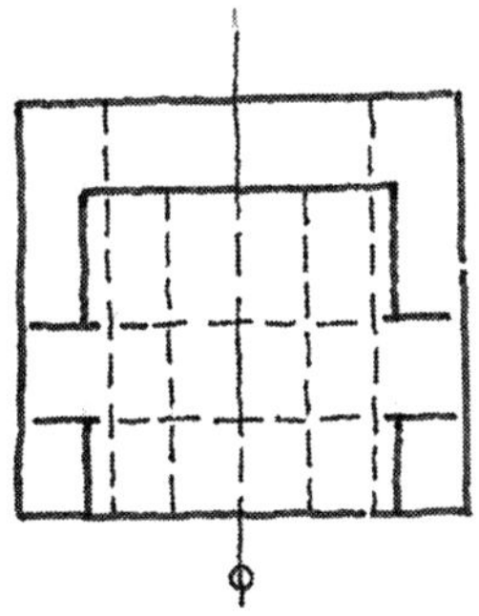

Drawn By:		FREEHAND SKETCHING	
	Scale	TITLE SECTIONING	
Date	Sheet of	Drawing Number	A 8.0

SKETCH THE SECTION AS CALLED FOR BY THE CUTTING PLANE.

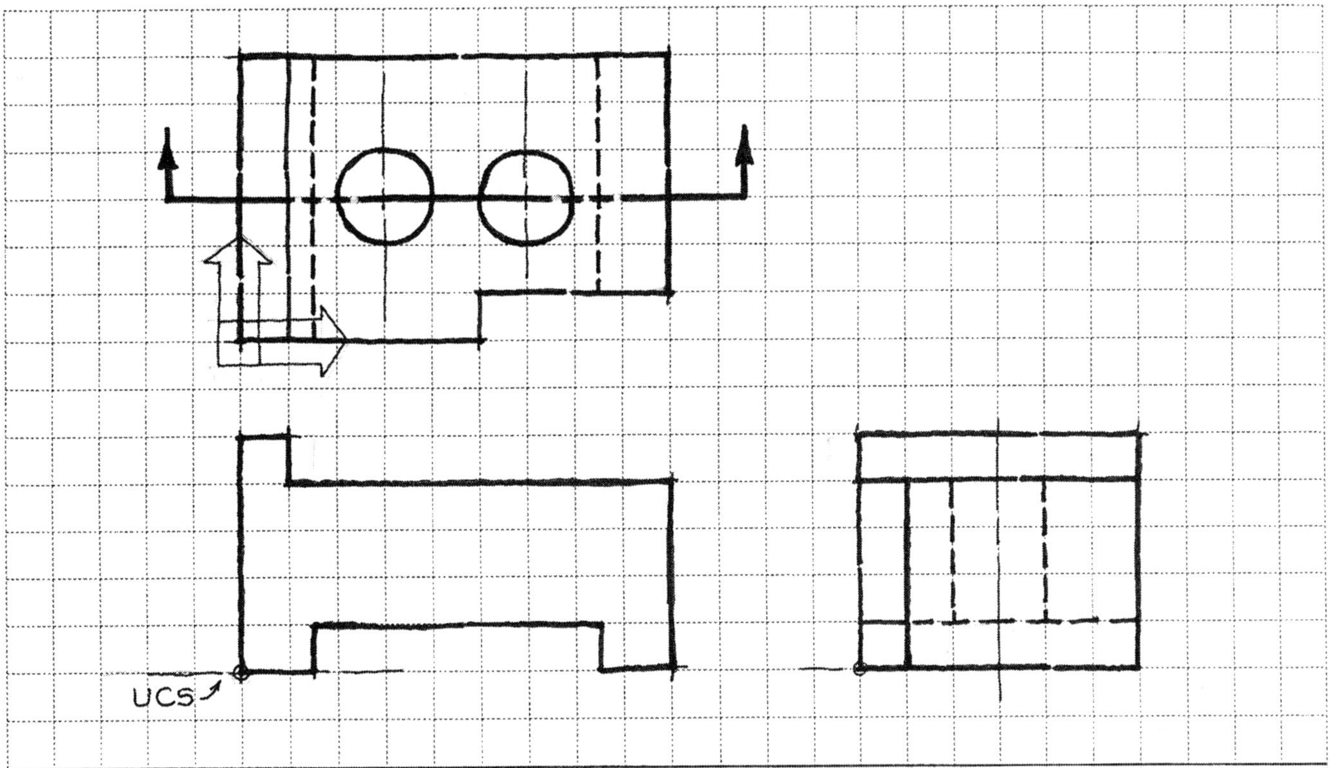

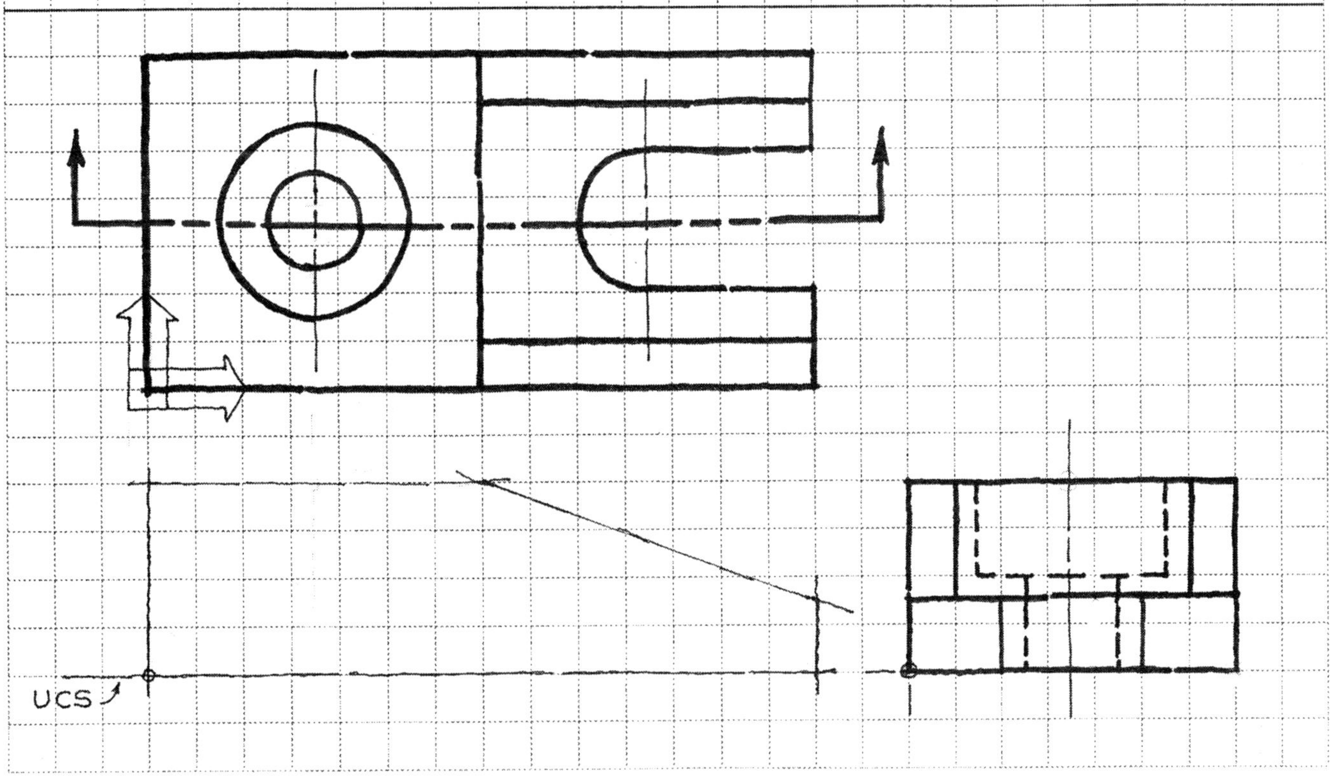

Drawn By:		FREEHAND SKETCHING	
	Scale	TITLE SECTIONING	
Date	Sheet of	Drawing Number A	8.1

HALF-SECTION: SKETCH THE RIGHT SIDE VIEW AS CALLED FOR BY THE CUTTING PLANE. OMIT HIDDEN LINES.

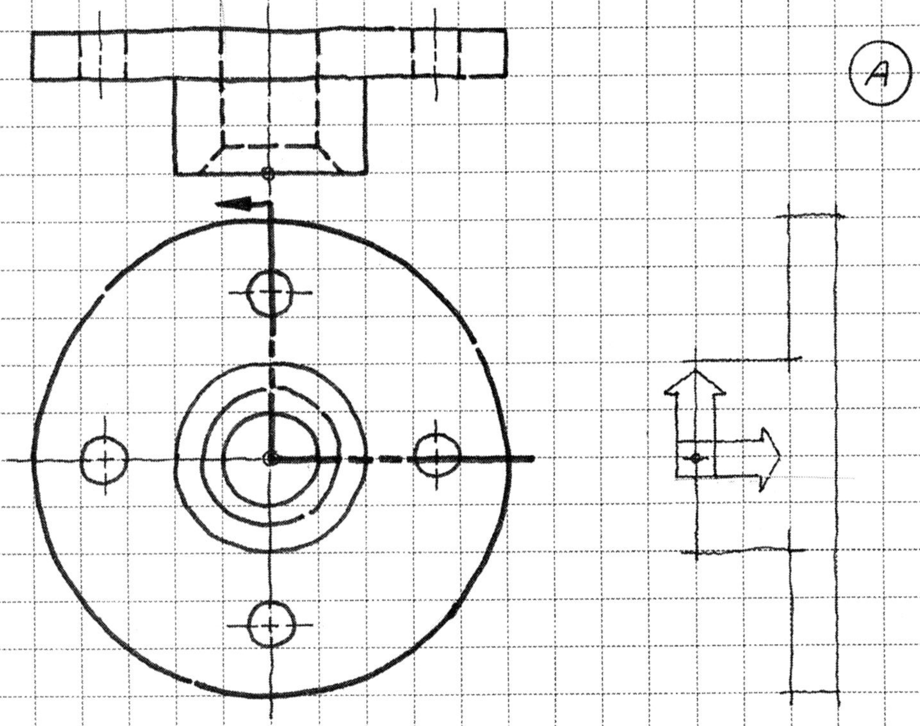

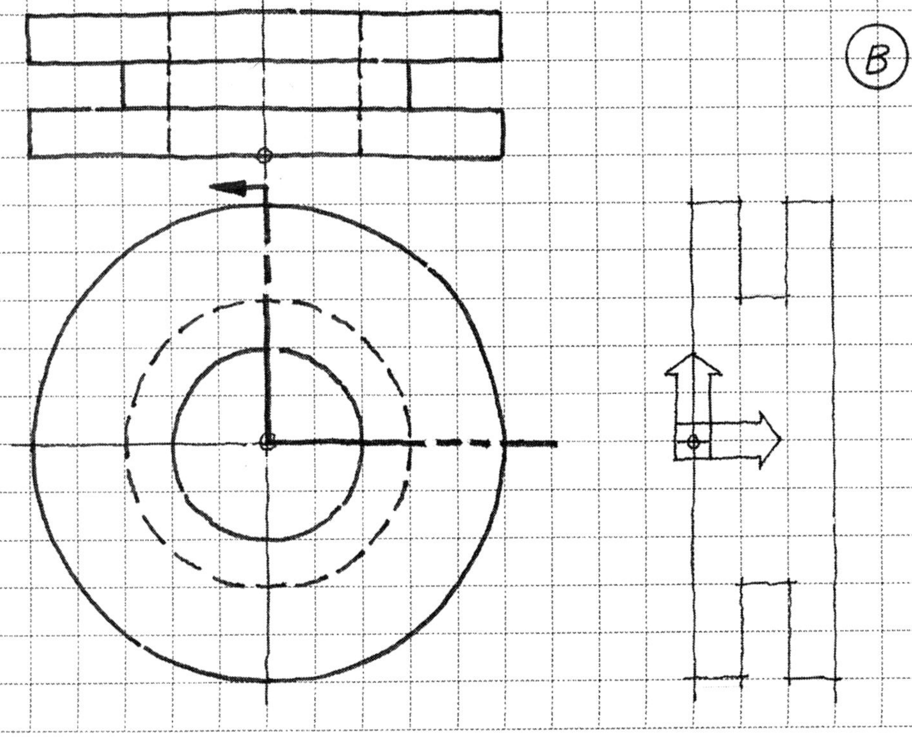

Drawn By:		FREEHAND SKETCHING	
	Scale	TITLE SECTIONING	
Date	Sheet of	Drawing Number A	8.1.1

ADD THE MISSING CUTTING PLANE AND COMPLETE THE MISSING VIEW OF EACH OBJECT IN FULL SECTION

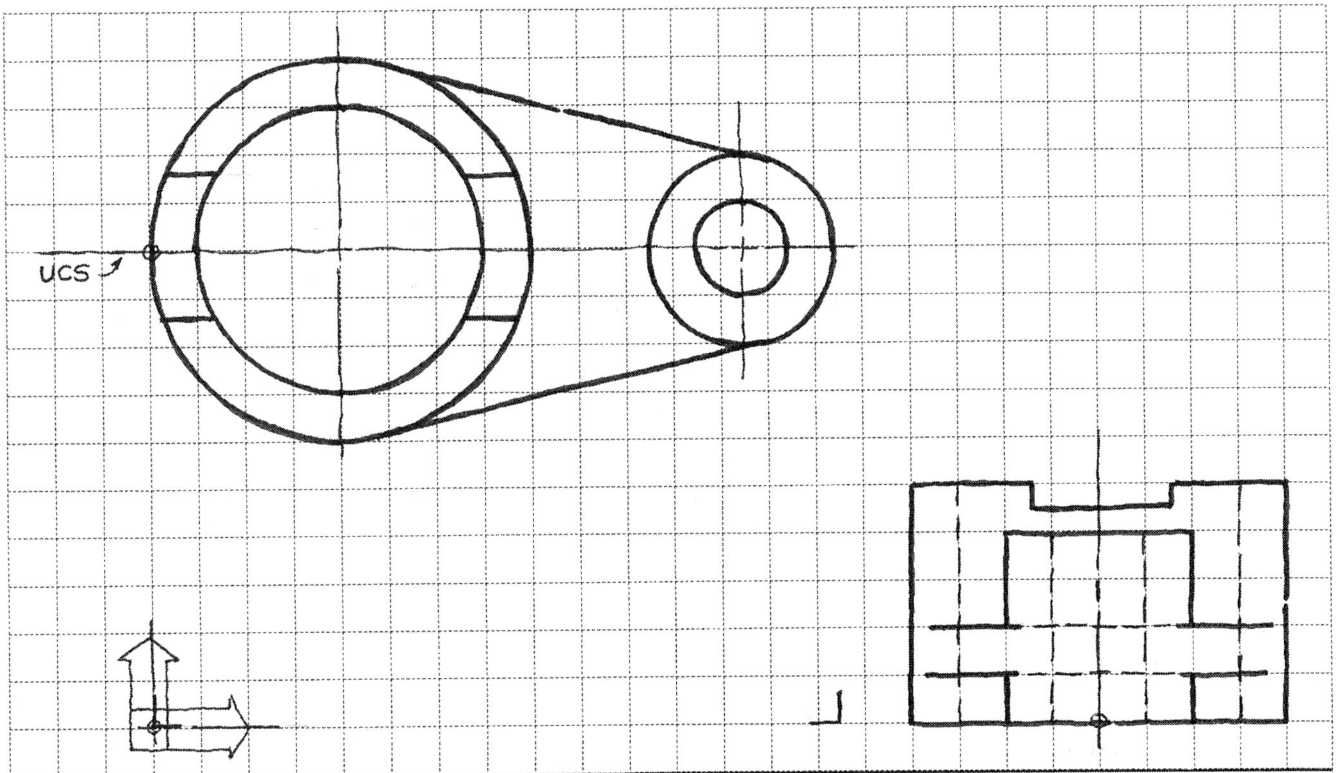

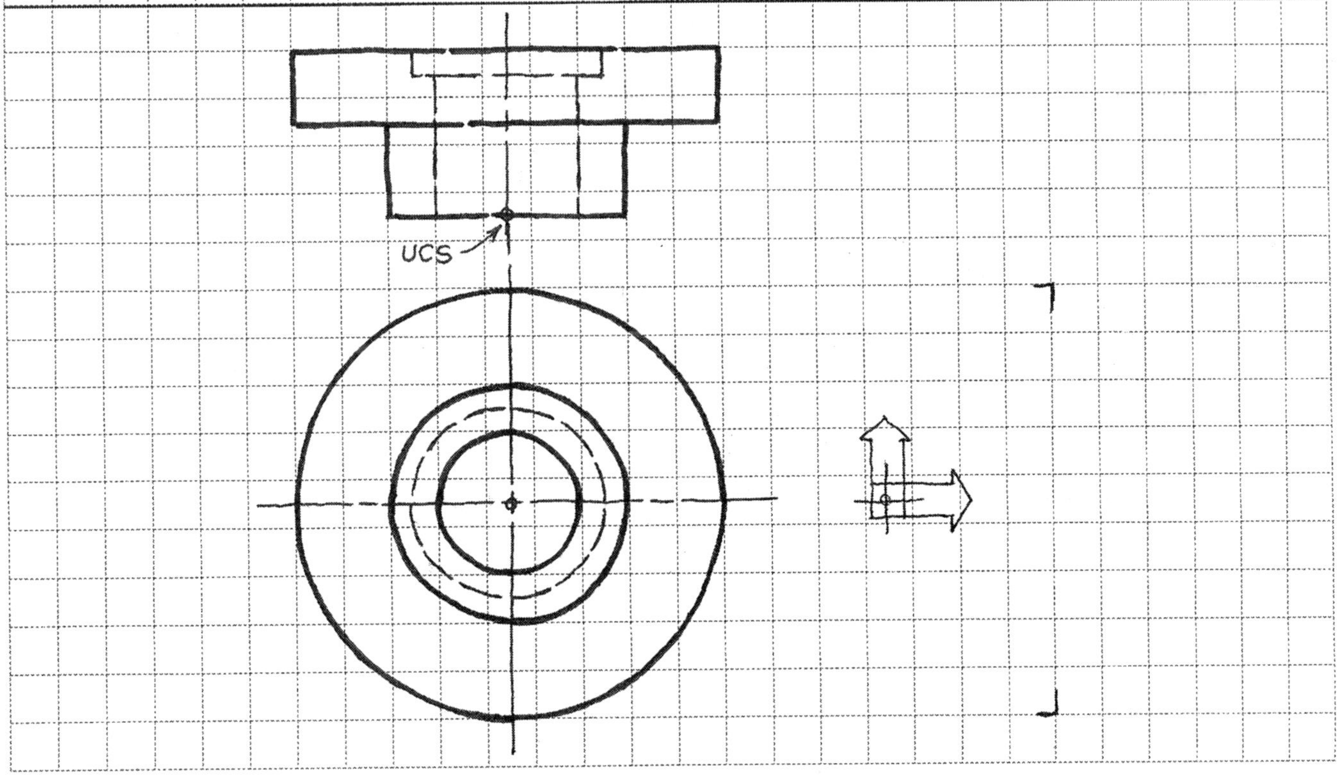

Drawn By:		FREEHAND SKETCHING	
	Scale	TITLE SECTIONING	
Date	Sheet of	Drawing Number A	8.2

SKETCH THE SECTION AS CALLED FOR BY THE CUTTING PLANE. USE STANDARD CONVENTIONS FOR REVOLVING OF RADIAL FEATURES AND THE TREATMENT OF RIBS.

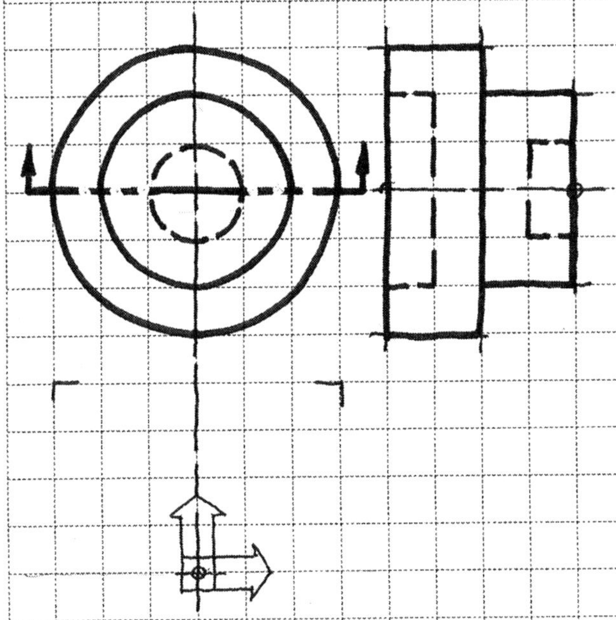

FULL SECTION

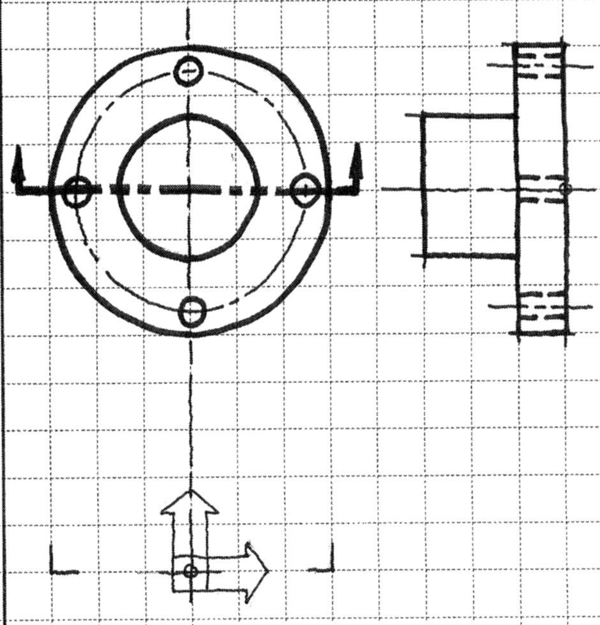

HALF SECTION

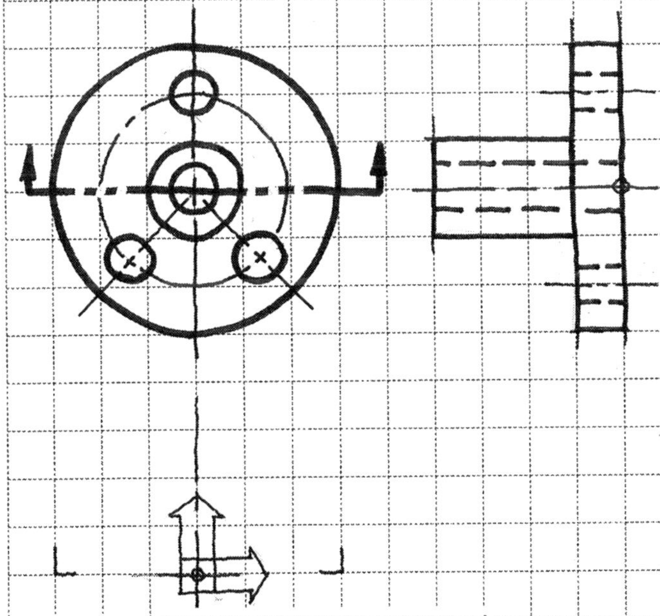

RADIAL FEATURE

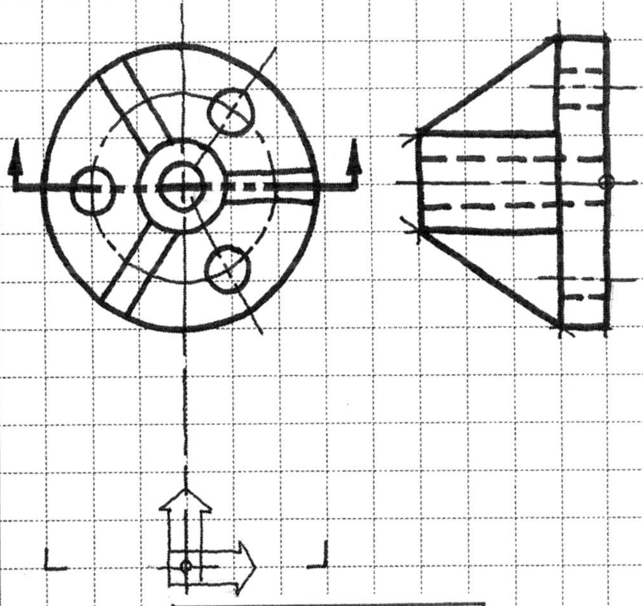

RADIAL FEATURE & RIBS

Drawn By:		FREEHAND SKETCHING	
	Scale	TITLE SECTIONING	
Date	Sheet of	Drawing Number A	8.3

PERFORM THE SECTION. USE STANDARD TREATMENT FOR RIBS AND RADIAL FEATURES.

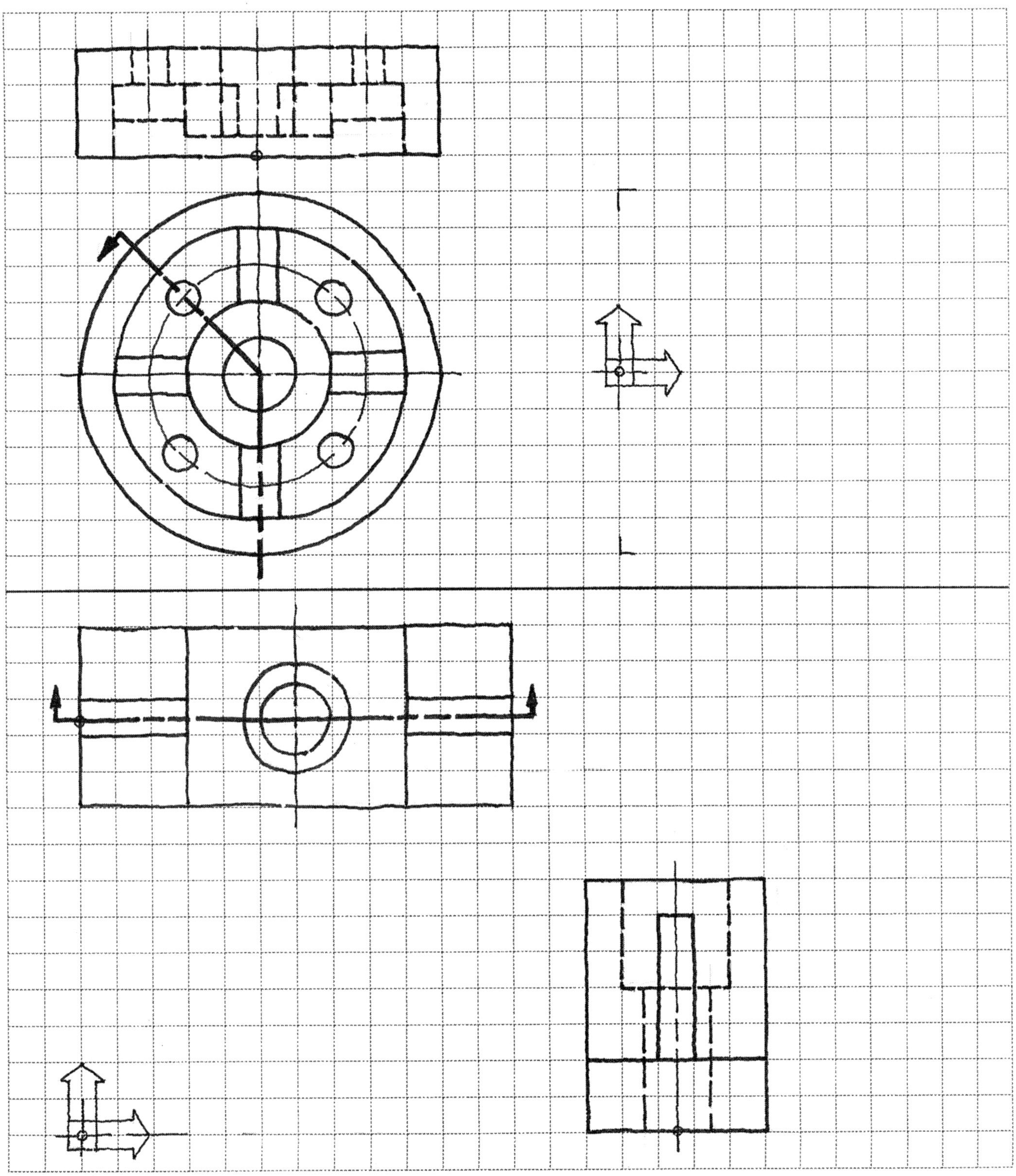

Drawn By:		FREEHAND SKETCHING	
	Scale	TITLE SECTIONING	
Date	Sheet of	Drawing Number A	8.3.1

SKETCH THE SECTION AS CALLED FOR BY THE CUTTING PLANE. USE STANDARD CONVENTIONS FOR TREATMENT OF SPOKES IN SECTION.

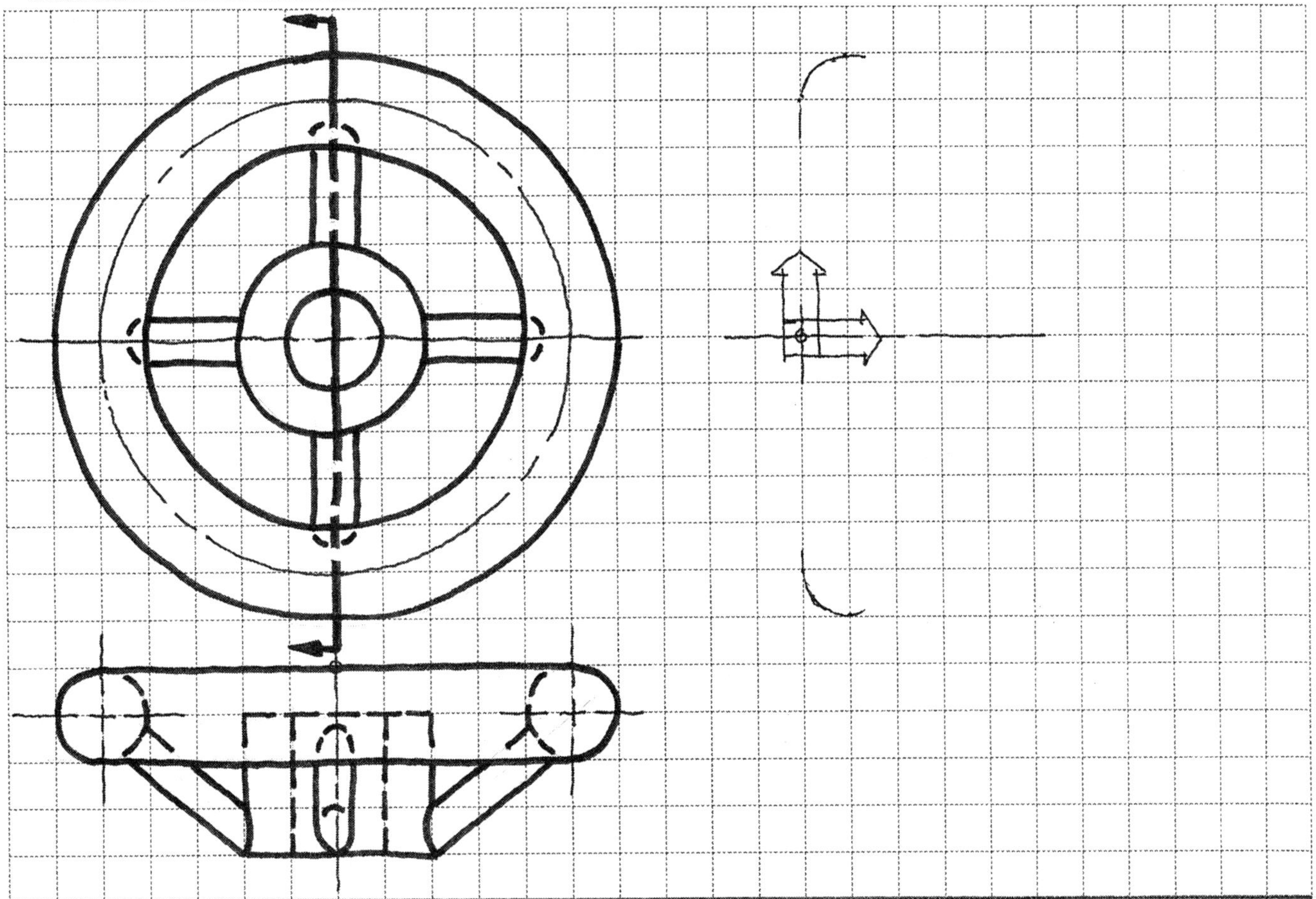

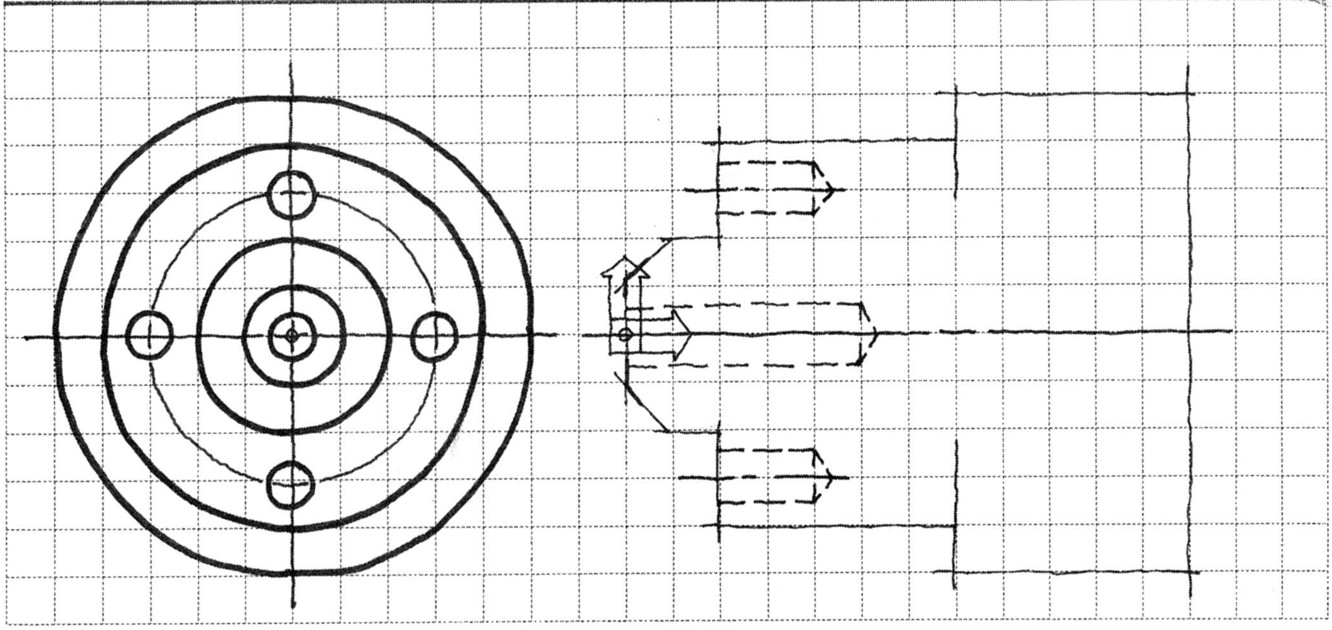

COMPLETE BROKEN SECTIONS AS REQUIRED.

Drawn By:		FREEHAND SKETCHING	
	Scale	TITLE SECTIONING	
Date	Sheet of	Drawing Number A	8.4

SKETCH THE REMOVED SECTIONS AS DESCRIBED BY THE CUTTING PLANES. PLACE VIEWS IN THE POSITIONS MARKED.

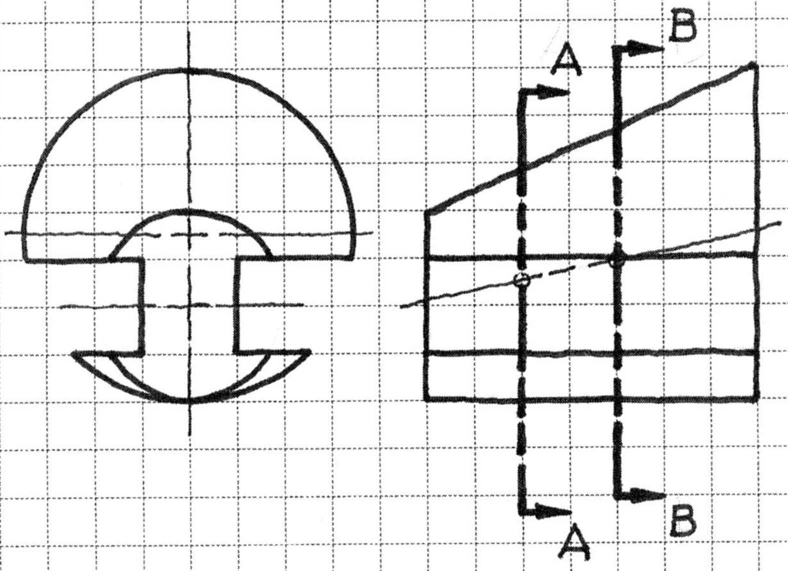

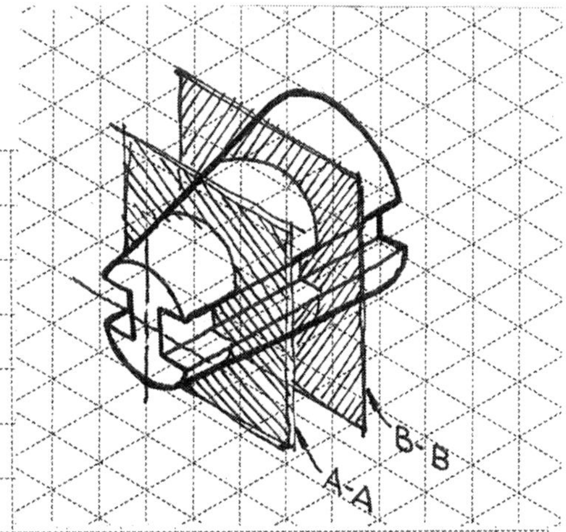

MOVE UCS AS REQUIRED

SECTION A-A
SCALE 2:1

SECTION B-B
SCALE 2:1

Drawn By:		FREEHAND SKETCHING	
	Scale	TITLE SECTIONING	
Date	Sheet of	Drawing Number A	8.5

SKETCH THE REMOVED SECTIONS IN THE SPACE PROVIDED.

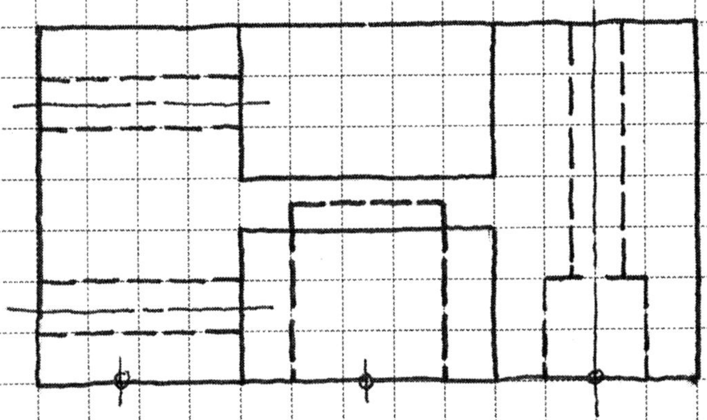

MOVE UCS AS REQUIRED

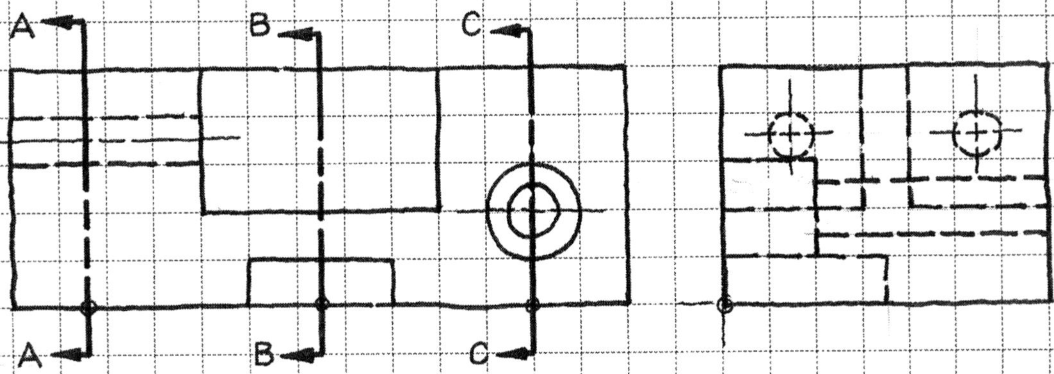

SECTION A-A

SECTION B-B

SECTION C-C

Drawn By:		FREEHAND SKETCHING	
	Scale	TITLE SECTIONING	
Date	Sheet of	Drawing Number A	8.5.1

PRODUCE THE REQUIRED REVOLVED SECTION.

SQUARE PRISM

CIRCULAR CYLINDER

TRIANGULAR PRISM

"I" MEMBER

"T" MEMBER

PRODUCE THE SPECIFIED MATERIAL SYMBOL.

CAST IRON

STEEL

ALUMINUM

BRONZE

ZINC

SHOW CONVENTIONAL BREAK TREATMENT

PLATE: SHORT BREAK

PLATE: LONG BREAK

S-BREAK: SOLID ROD

S-BREAK: HOLLOW CYLINDER

Drawn By:		FREEHAND SKETCHING	
	Scale	TITLE SECTIONING	
Date	Sheet of	Drawing Number A	8.6

IN EACH PROBLEM, DETERMINE THE CORRECT TREATMENT FOR THE SECTIONED FEATURES.

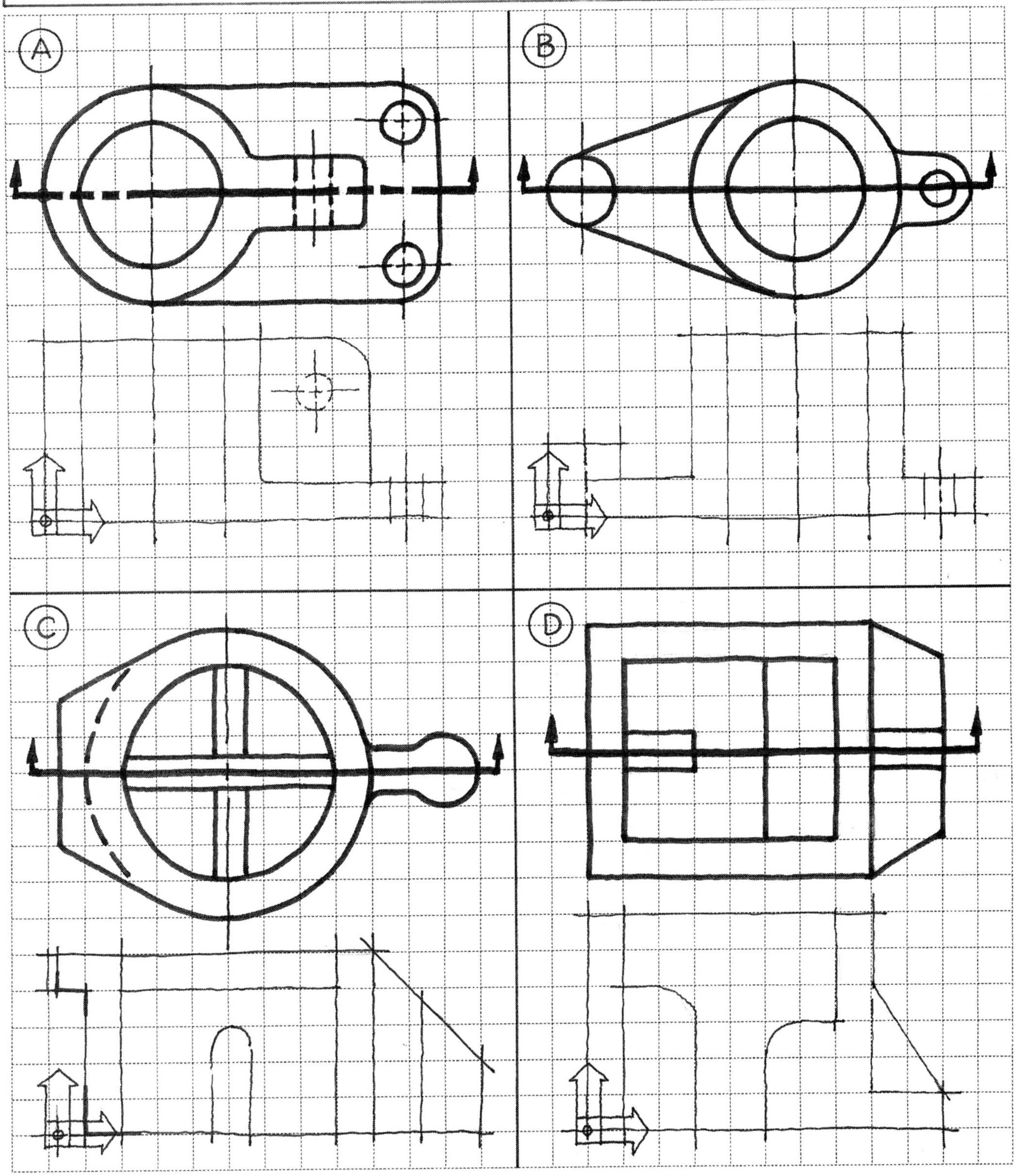

Drawn By:		FREEHAND SKETCHING	
	Scale	TITLE SECTIONING	
Date	Sheet of	Drawing Number A	8.7

SHOW POSITION OF OFFSET CUTTING PLANE IN PROBLEMS A & B.
COMPLETE FRONT VIEW IN PROBLEM C AS A FULL OFFSET SECTION.

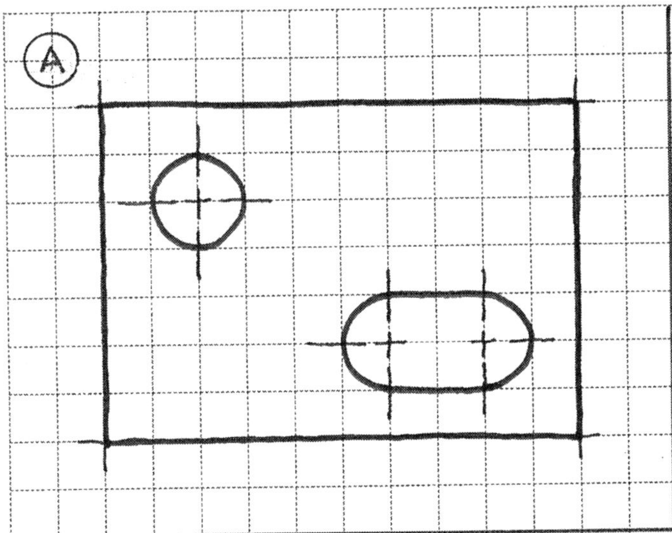

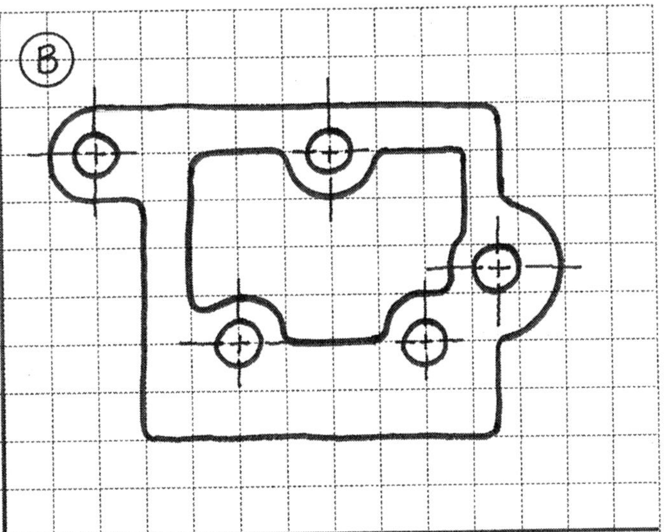

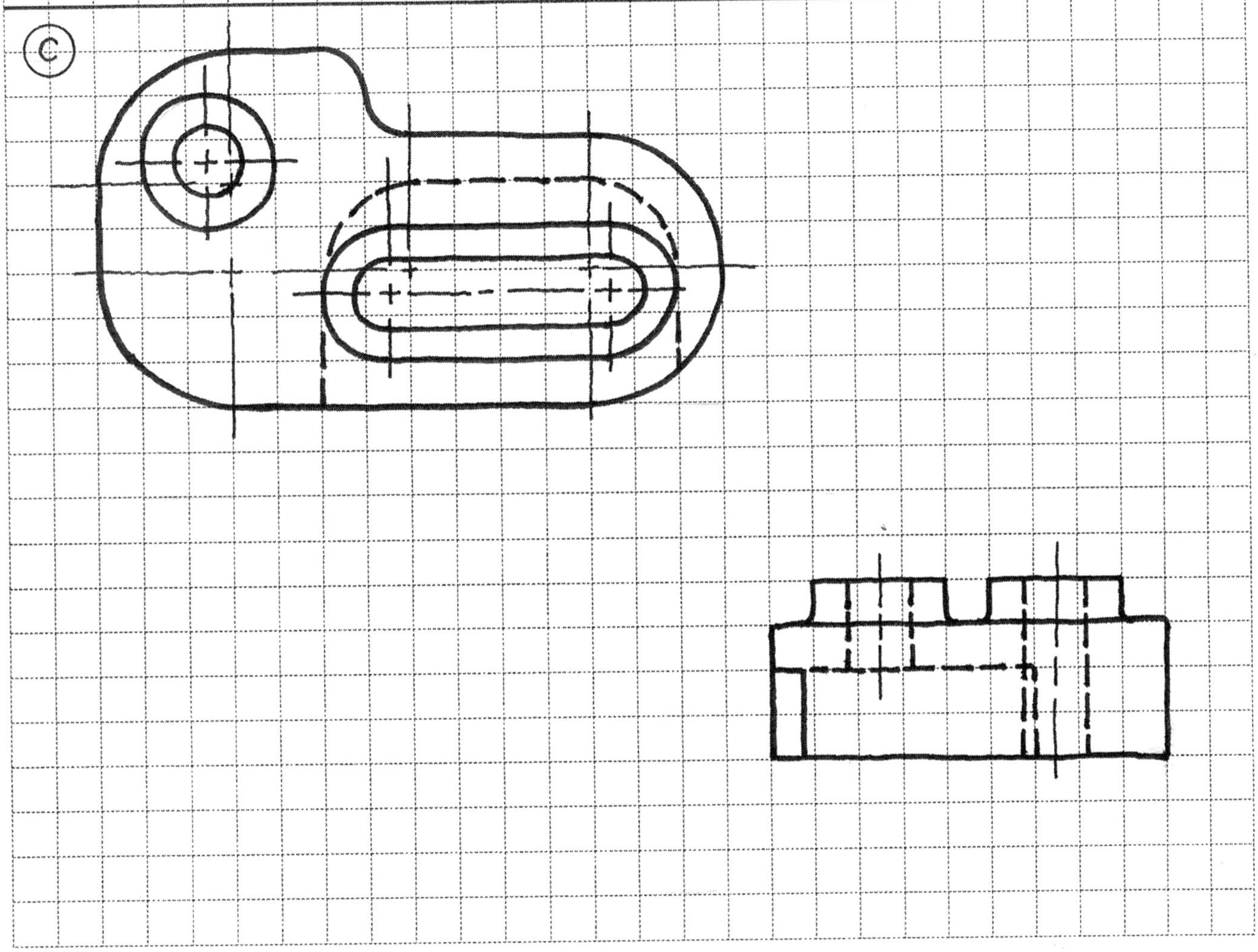

Drawn By:		FREEHAND SKETCHING	
	Scale	TITLE SECTIONING	
Date	Sheet of	Drawing Number A	8.8

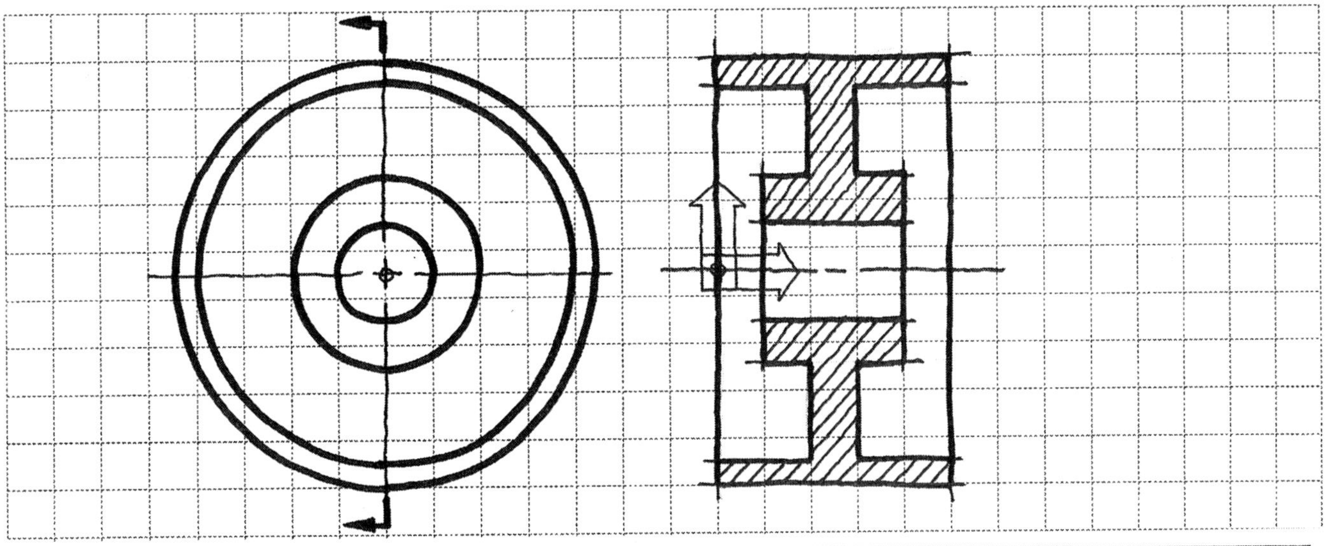

GIVEN: FRONT AND RIGHT SIDE (FULL SECTION) VIEWS OF A GEAR BLANK.
COMPLETE AN ISOMETRIC SKETCH IN HALF-SECTION.

Drawn By:		FREEHAND SKETCHING	
	Scale	TITLE SECTIONING	
Date	Sheet of	Drawing Number A	8.9

SHOW THE INCOMPLETE VIEW AS A SECTIONAL ASSEMBLY.

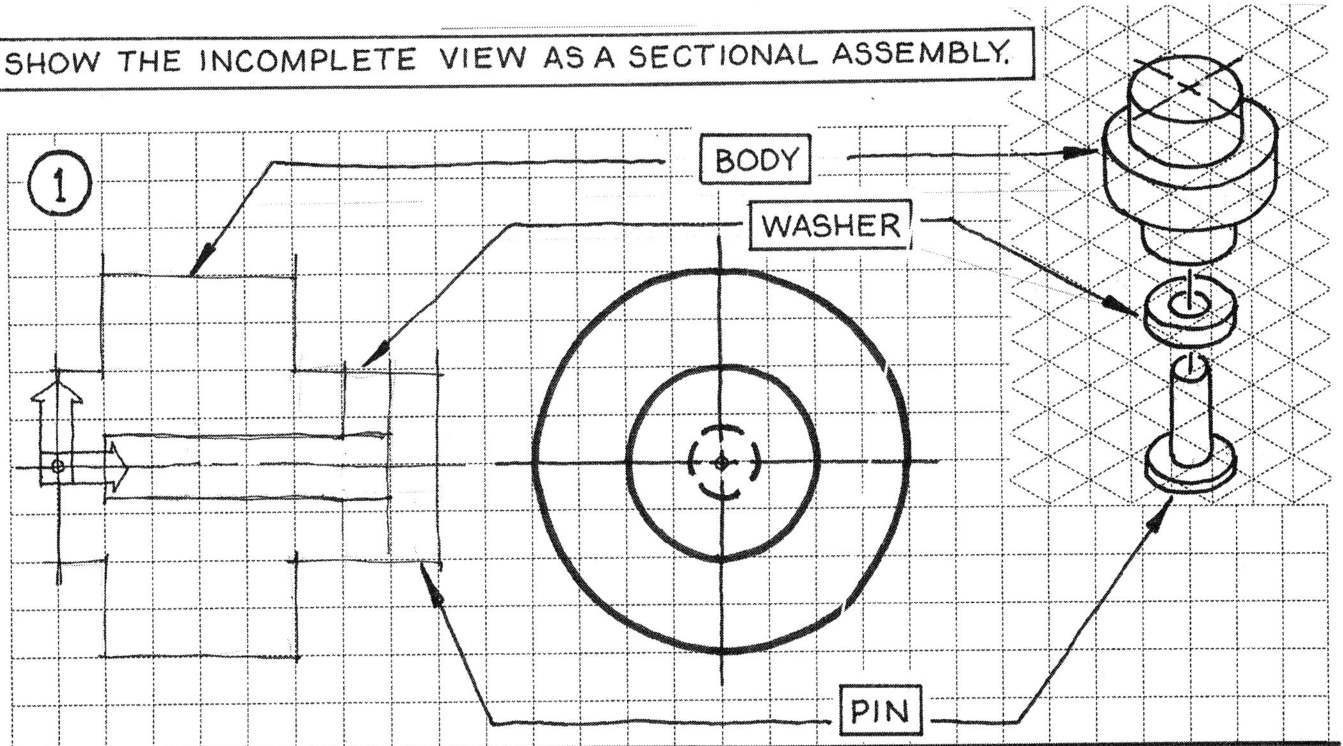

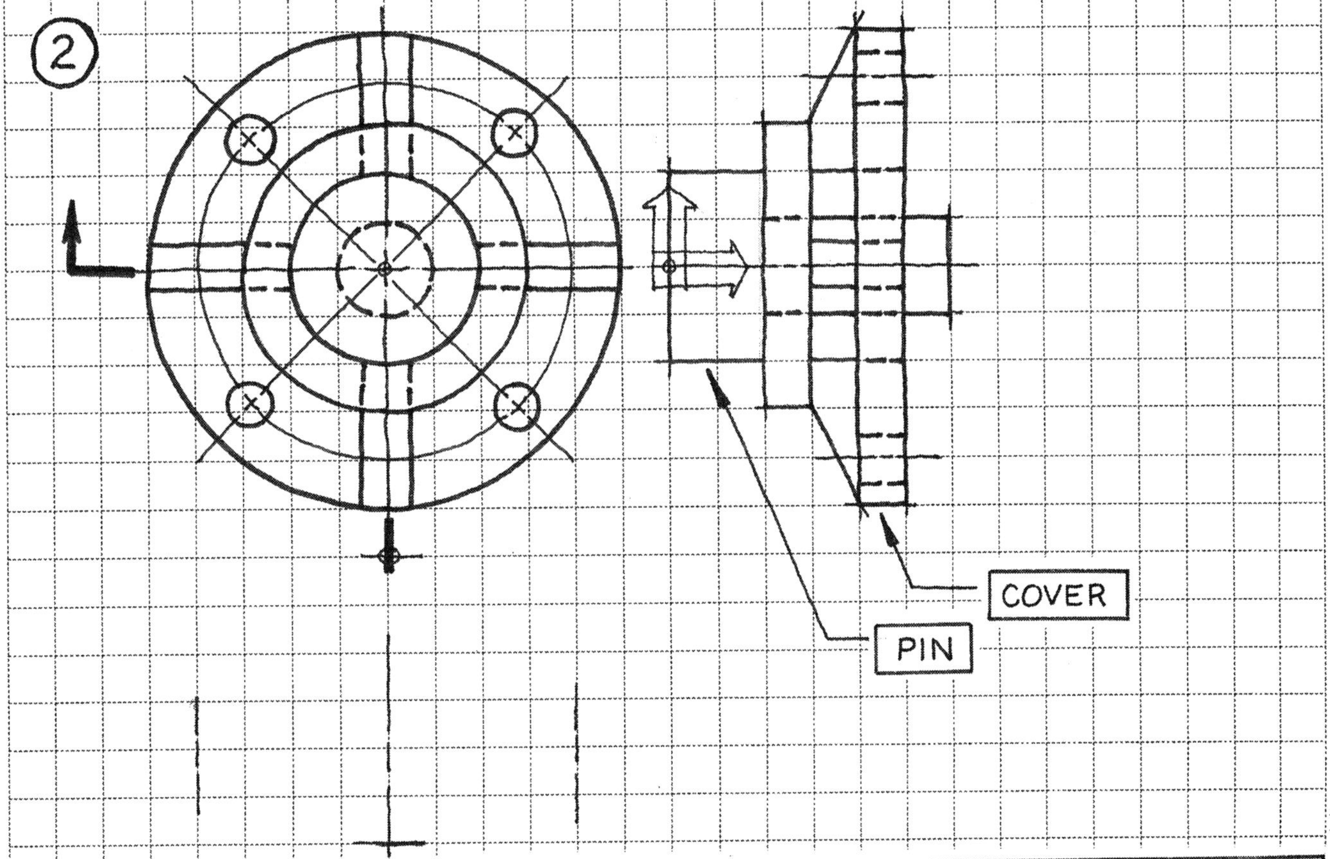

COMPLETE THE HALF-SECTION OF THE ASSEMBLY.

Drawn By:		FREEHAND SKETCHING	
	Scale	TITLE SECTIONING	
Date	Sheet of	Drawing Number A	8.10

THE FULL-SIZE DETAILS PRESENTED ON THIS PAGE ARE TO BE USED TO COMPLETE THE SECTIONAL ASSEMBLY ON SHEET 8.12.

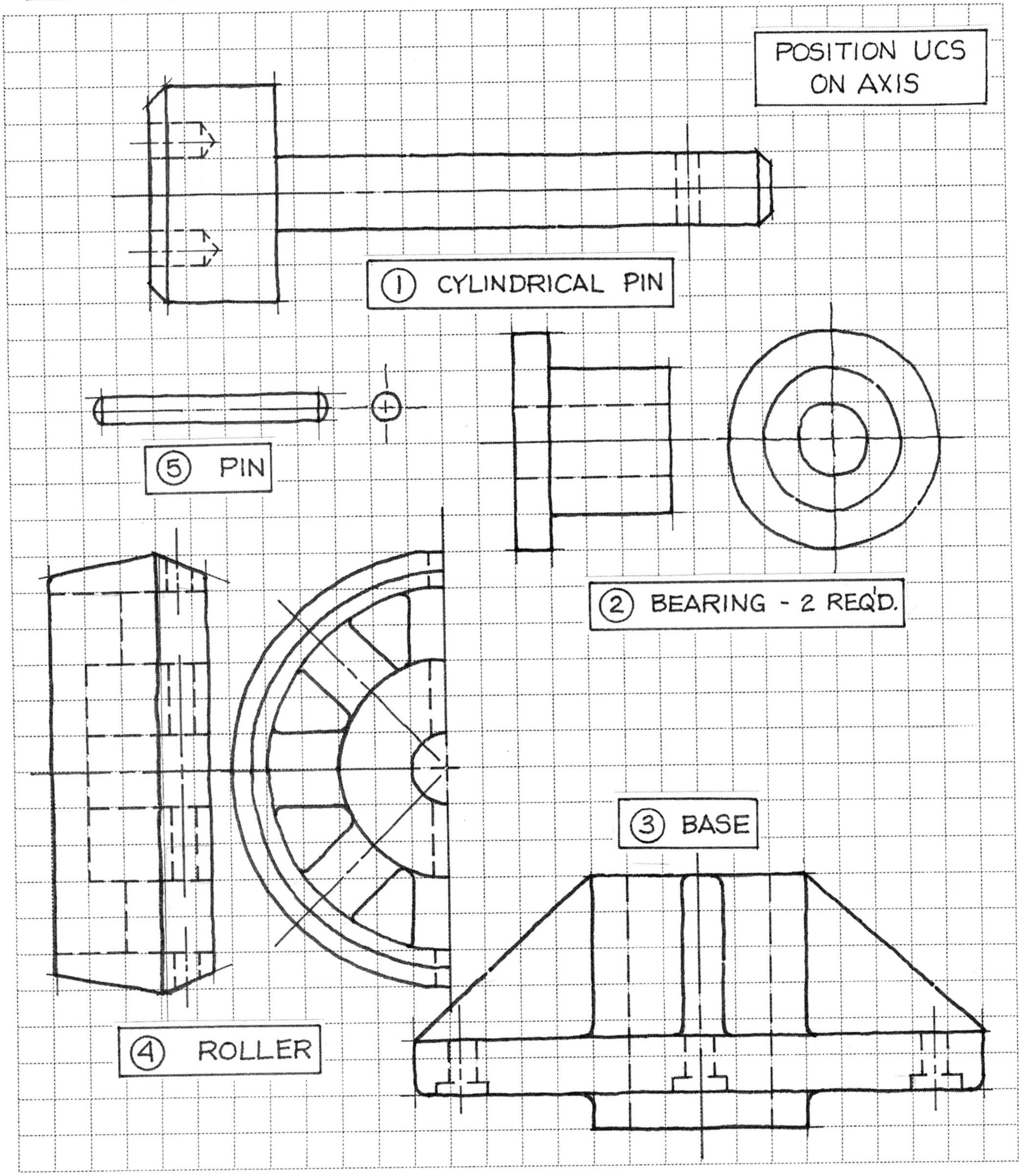

Drawn By:		FREEHAND SKETCHING	
	Scale 1:1	TITLE DETAILS	
Date	Sheet 1 of 2	Drawing Number A	8.11

USING THE DETAILS ON SHEET 8.11, COMPLETE A FULL SECTIONAL ASSEMBLY DRAWING IN THE SPACE PROVIDED. NOTE THAT THE PARTS HAVE BEEN IDENTIFIED.

POSITION UCS ON AXIS

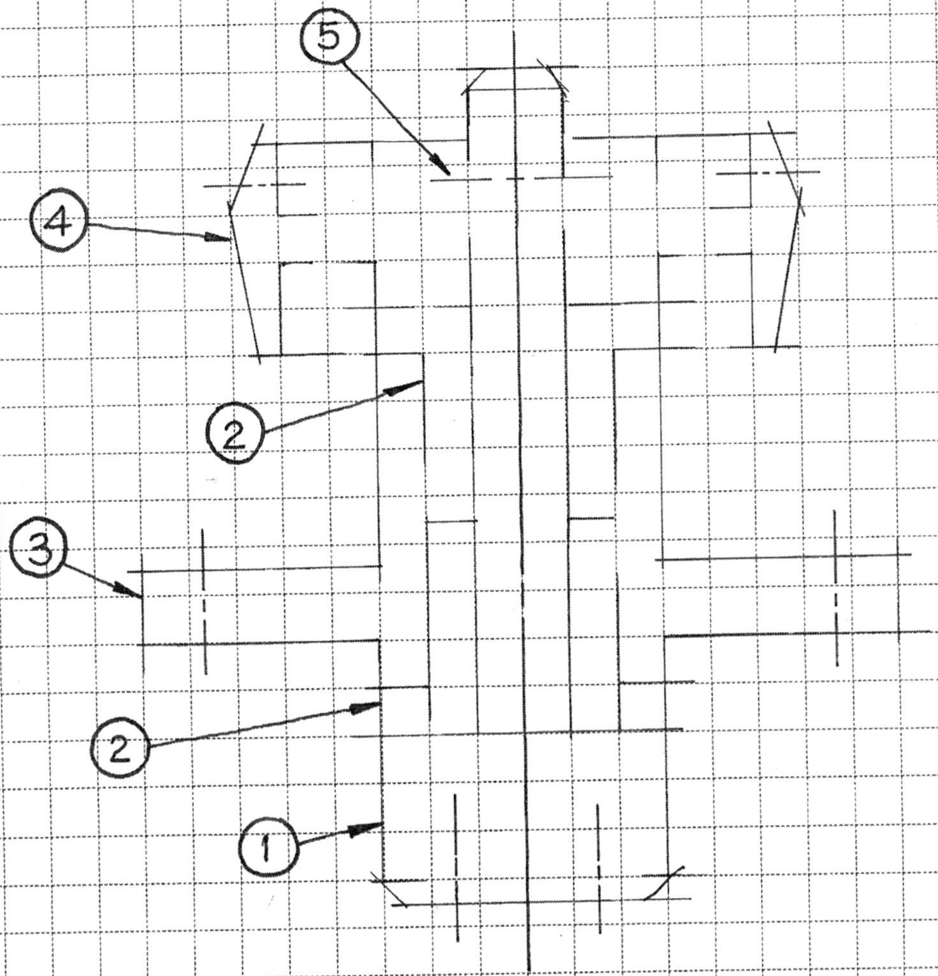

Drawn By:		FREEHAND SKETCHING	
	Scale 1:1	TITLE ASSEMBLY	
Date	Sheet 2 of 2	Drawing Number A	8.12

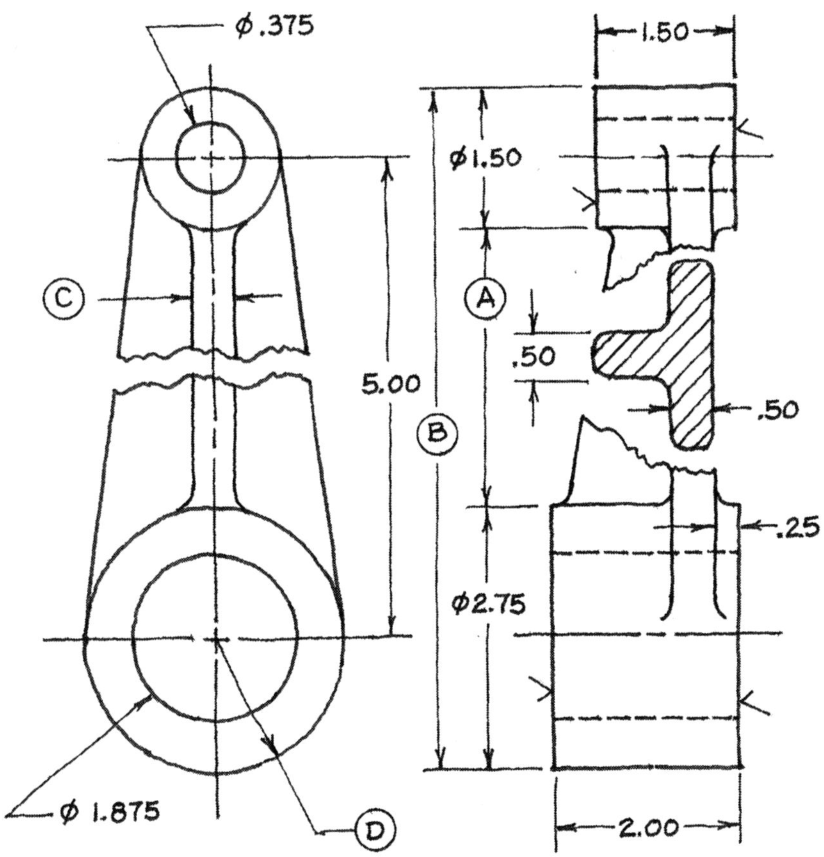

SHAFT HANGER

MATL – CI, 4 REQD

NOTE: UNLESS OTHERWISE SPECIFIED –
ROUNDS AND FILLETS .08R
TOLERANCE ON TWO-DECIMAL DIMENSIONS ± .02
TOLERANCE ON THREE-DECIMAL DIMENSIONS ± .004

QUESTIONS:

1. How many surfaces on the casting require finishing?
2. What type of section is used?
3. What is the minimum allowable center-to-center distance between holes?
4. What is the maximum allowable height of the part at (B)?
5. What are the maximum and minimum allowable dimensions for (A) and (C)?
6. What type of break line is used on the sectional view?
7. What is the maximum permissable wall thickness at the large hole?
8. What are the nominal dimensions for (A), (B), (C), and (D)?

ANSWERS:

1. ____________
2. ____________
3. ____________
4. ____________
5. ____________
6. ____________
7. ____________
8. ____________

Drawn By:		FREEHAND SKETCHING	
	Scale 2 : 1	TITLE READING DRAWINGS	
Date	Sheet of	Drawing Number A	8.13

9 Dimensioning Practices

Coordinated with VMGED Chapter 15

Dimensioning—General Conditions

The full description of an object includes its shape, its size, and any and all information necessary for it to be manufactured. In previous chapters you gained experience in describing the appearance of an object. In this chapter you will learn the basics of describing its size.

Most sketches are made *not to scale* (NTS). This means that the appearance is proportional, at best. The numbers associated with the sketch represent actual sizes. This is shown in Fig. 9-1. Because sketches are visual approximations, correct dimensions are very important.

A dimension is recorded according to a set of specific rules. In its most basic form, a dimension has two components:

1. The numerical size (size dimension).

2. The position of the feature (location dimension).

A correctly dimensioned drawing contains size and location information necessary and sufficient for it to be produced within an acceptable range of variability. Figure 9-2 shows the necessary and sufficient information to locate and size a hole on a plate.

Dimension Symbology

Dimensions can be linear, radial, angular, and diametrical. Figure 9-3 shows the components of these basic dimension types. Standard industrial practice is to use a textual note whenever possible.

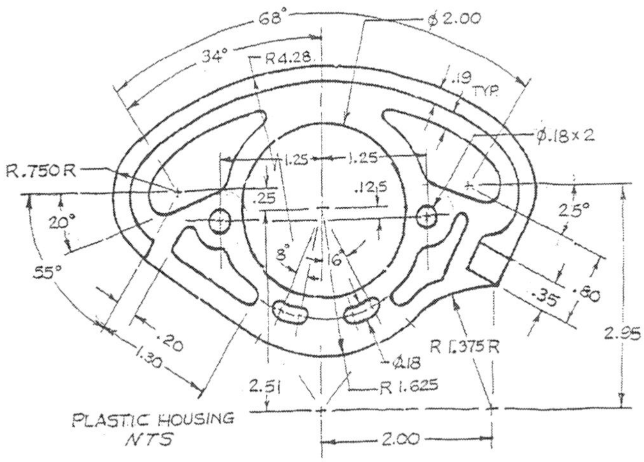

Fig. 9-1. A proportional sketch with dimensional information.

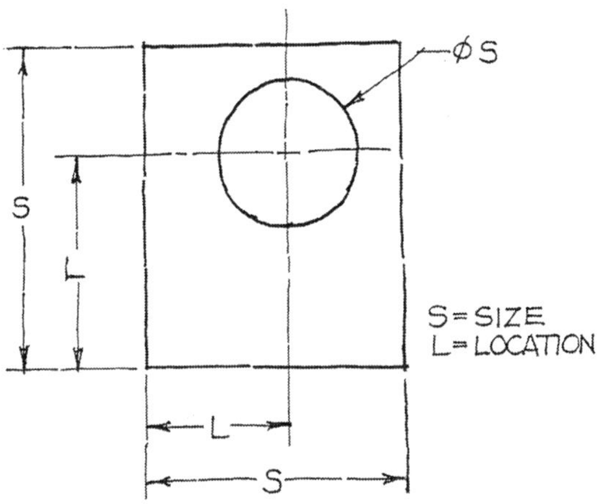

Fig. 9-2. Size and location dimensions.

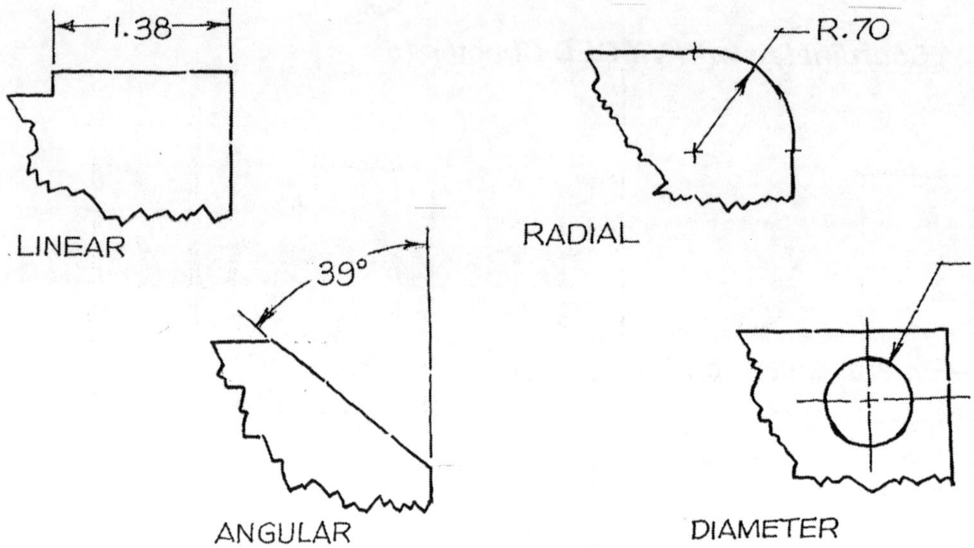

Fig. 9-3. Dimensioning symbology.

Sheet 9.0 (page 205) outlines the steps you should follow in dimensioning an object. Spend some time with this sheet before starting the exercises. Note in Step 2 that holes are dimensioned by their diameters and curved corners are dimensioned by their radii. Step 4 shows that a dimension can function both as size and location. Step 6 shows a general note. Also, in this final step, dimensions—called datums—are associated with a common base line.

Dimensioning Geometric Primitives

All parts, no matter how sophisticated, can be broken into primitive shapes. Correct dimensioning practice is then a process of correctly sizing and locating features within the overall structure of the object. Figure 9-4 depicts the necessary size information for several primitive shapes. Keep two points in mind:

1. Holes are dimensioned in the view in which the axis of the hole appears as a point (circle view of the hole).

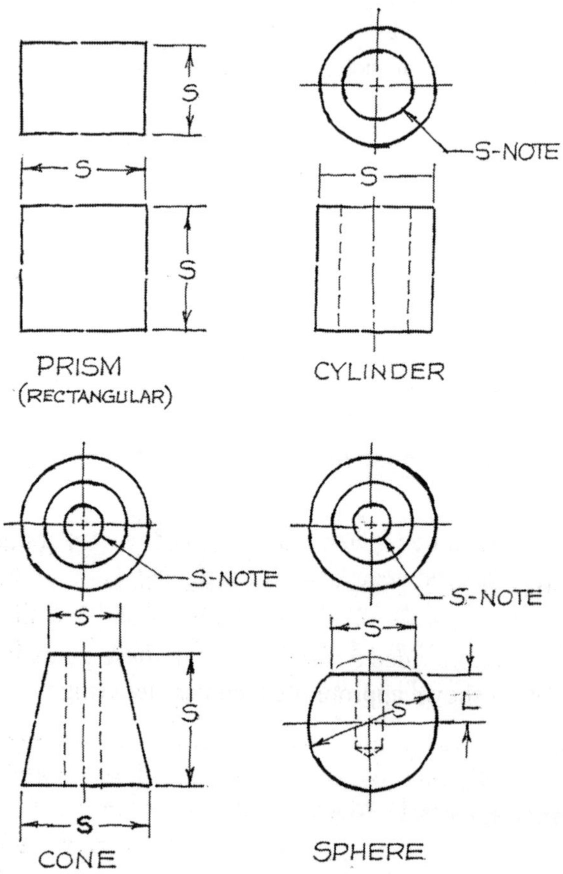

Fig. 9-4. Dimensions of primitives.

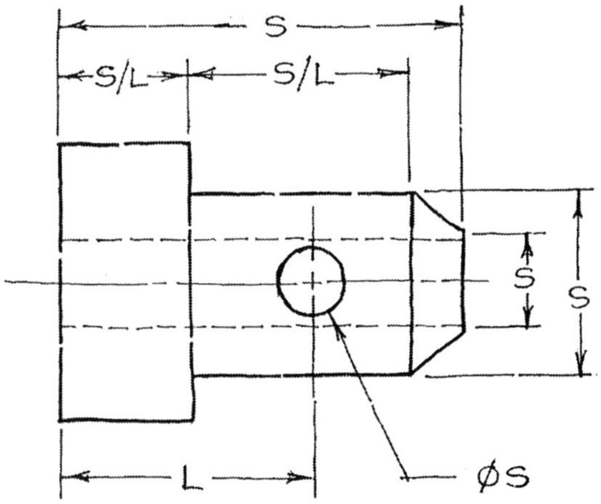

Fig. 9-5. A dimensioned composite object.

2. Cylindrical diameters are dimensioned in the view in which the axis of the cylinder appears normally (rectangular view of cylinder).

Figure 9-5 applies these techniques to composite geometry.

Technical Lettering

In a dimensioned sketch, legible lettering is of paramount importance. Never letter without using light guide lines. In the lab sheets you have been turning in, the small tick marks have provided a start and end to these guide lines. Figure 9-6 shows appropriate technical lettering.

Assignment: Complete a technical sketch of an object you can easily hold in your hand. Use ortho (square) grid paper. Using dimensioning techniques shown in Figs. 9-3 and 9-4, show overall height, width, and depth. Use appropriate technical lettering for the numerals.

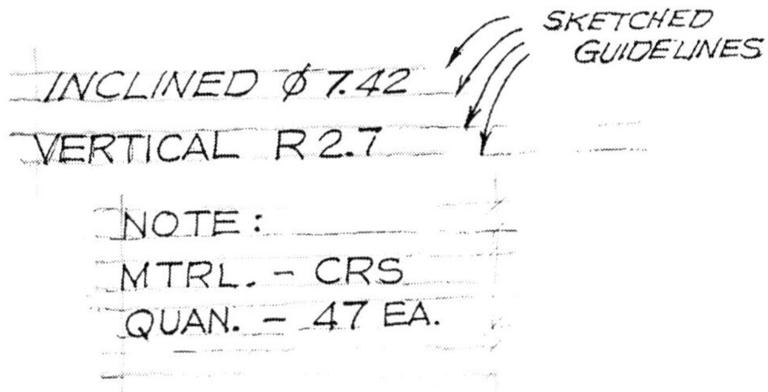

Fig. 9-6. Technical lettering.

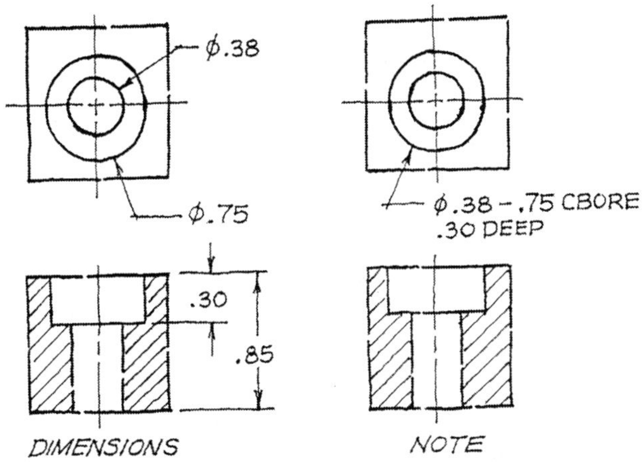

Fig. 9-7. Comparison of dimensions and notes.

Notes

Many times using a written technical note is clearer than inserting traditional dimensions. Figure 9-7 (previous page) shows a hole detail. The required size information has been recorded both in dimensions and notes. Which do you believe is more efficient?

CAD Example—2D

CAD drafting programs include semiautomatic dimensioning functions. This means the program provides all sizes (from its data base of drawing elements); you must determine which dimensions are necessary and where to place them. Two-dimensional CAD automates and standardizes the drawing of extension and dimension lines, arrowheads, and dimension text. CAD dimensions are consistent and readable.

CAD Example—3D

Three-dimensional CAD models do not normally need to be dimensioned. The numerical information associated with the model can be interpreted directly by numerically-controlled machine tools and processing equipment. If the geometric data need be communicated to a human machine operator or technician, the model can be dimensioned in space (Fig. 9-8) or projected onto a 2D construction plane and dimensioned traditionally (Fig 9-9) .

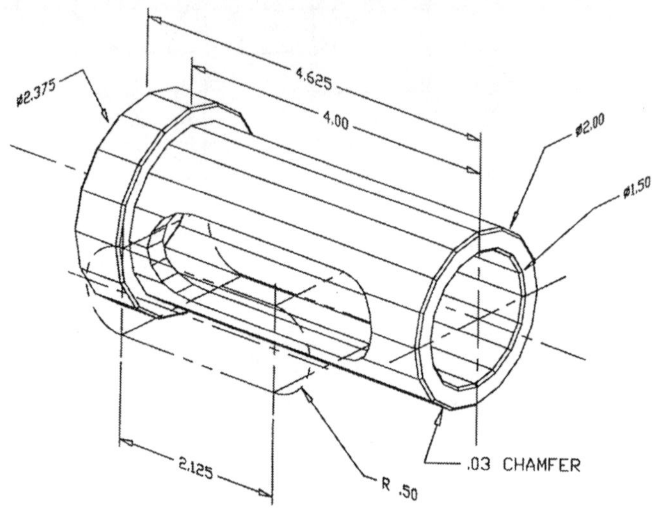

Fig. 9-8. An edited and dimensioned 3D model.

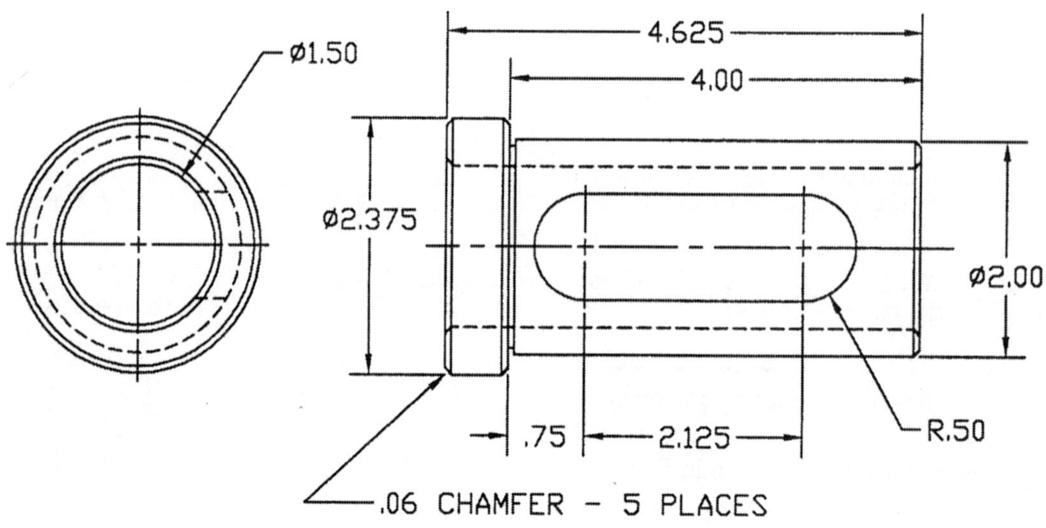

Fig. 9-9. Model projected into 2D and dimensioned.

FUNDAMENTALS OF DIMENSIONING

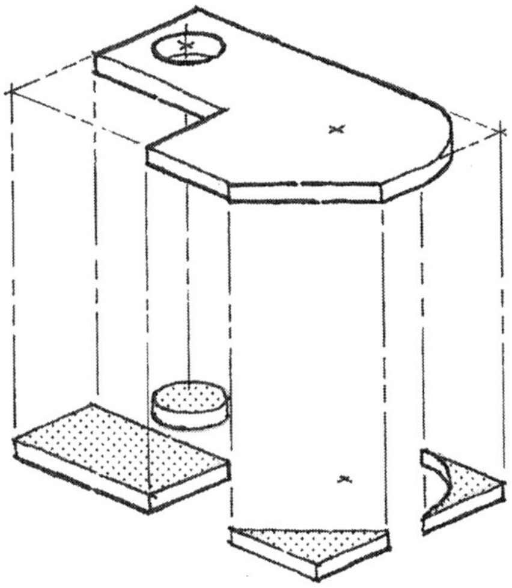

(1) DIVIDE OBJECT INTO COMPONENT GEOMETRIC SHAPES.

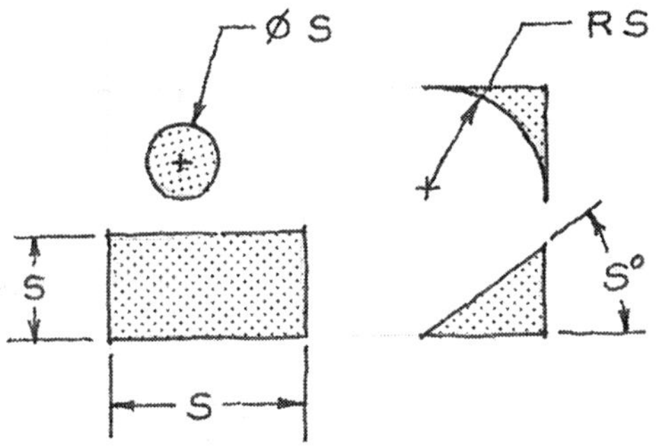

(2) DETERMINE SIZE DIMENSIONS FOR EACH COMPONENT SHAPE.

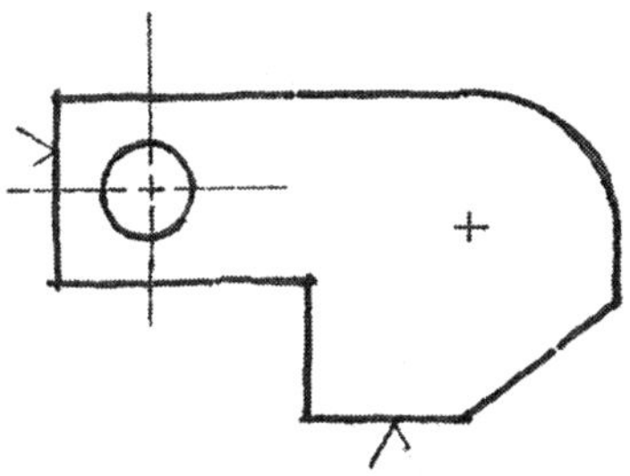

(3) IDENTIFY LOCATING CENTER LINES AND FINISHED SURFACES.

* ALSO SIZE DIMENSIONS

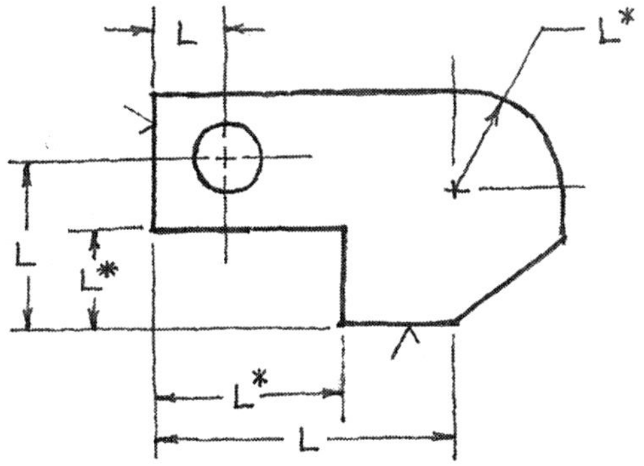

(4) DETERMINE LOCATION DIMENSIONS FOR EACH COMPONENT SHAPE.

D = .125 PLATE

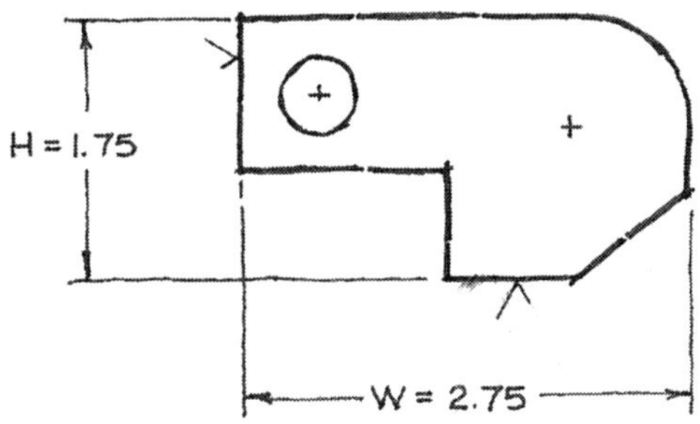

(5) SHOW OVERALL DIMENSIONS OF HEIGHT, WIDTH, AND DEPTH.

MTRL: .125 ALUM PL
TOLER.: +/– .01
QUAN.: 1700
PART No. NF89

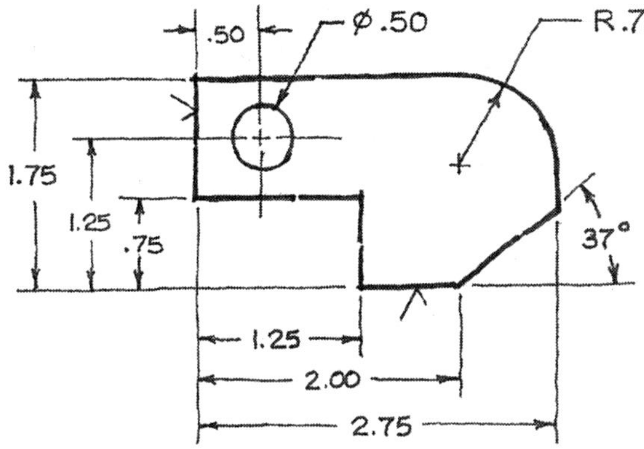

(6) ADD REQUIRED NOTES OR PROCEDURES.

Drawn By:		FREEHAND SKETCHING	
	Scale NTS	TITLE INFORMATION SHEET	
Date	Sheet of	Drawing Number A	9.0

GIVEN AN OBJECT AND EXTENSION LINES, ADD DIMENSION LINES AND TERMINATORS.

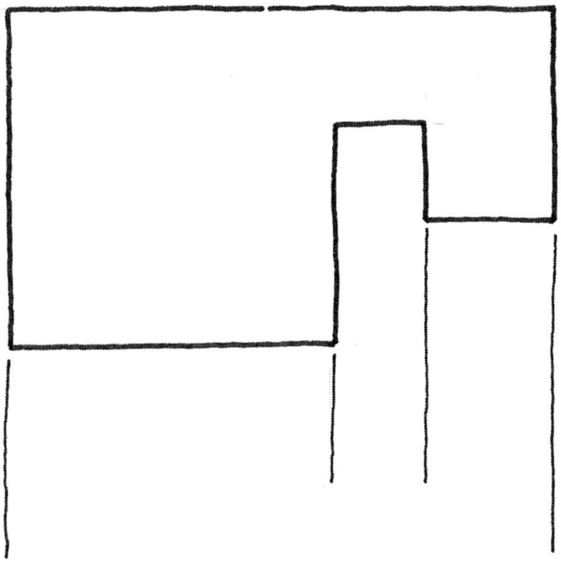

PLACE THE NUMERICAL VALUE 4.56 ON A COMPLETE DIMENSION.

DIMENSION THE .75 RADIUS ROUNDED CORNER AND .25 DIAMETER HOLE.

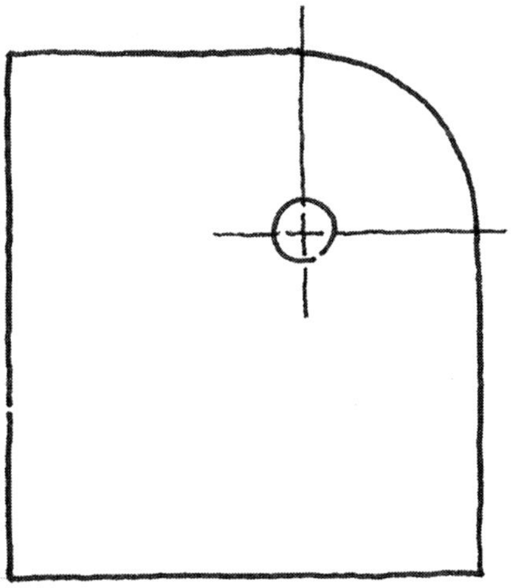

DIMENSION THE 3.5 DIAMETER CYLINDER AND .50 DIAMETER HOLE.

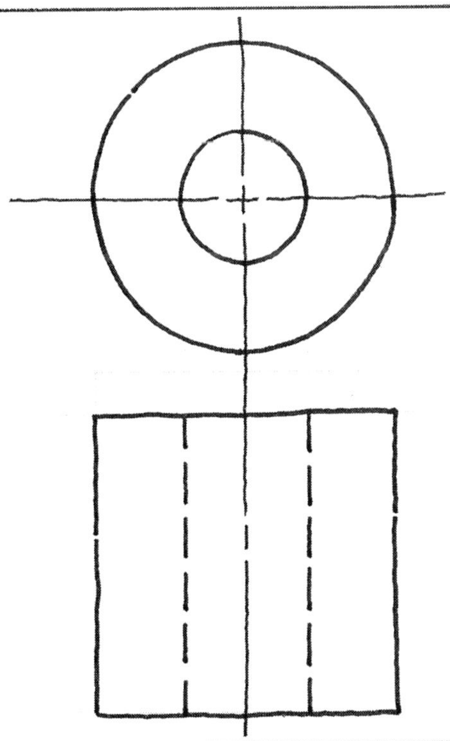

Drawn By:		FREEHAND SKETCHING	
	Scale N/A	TITLE DIMENSIONING	
Date	Sheet of	Drawing Number A	9.1

PIPE DATA: 1.50 O.D. / 1.00 I.D. / 2.25 LONG. USE UNIDIRECTIONAL DIMENSIONING METHOD.

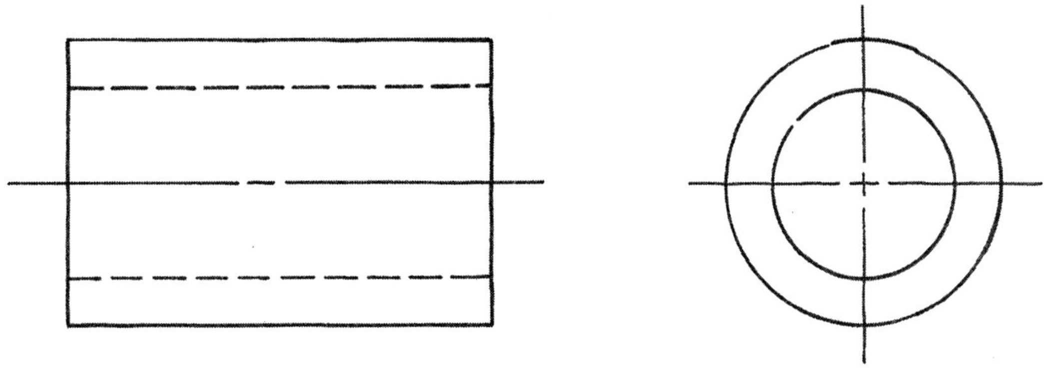

DIMENSION THE CYLINDRICAL PLUG USING COMMON FRACTIONS. USE THE ALIGNED METHOD.

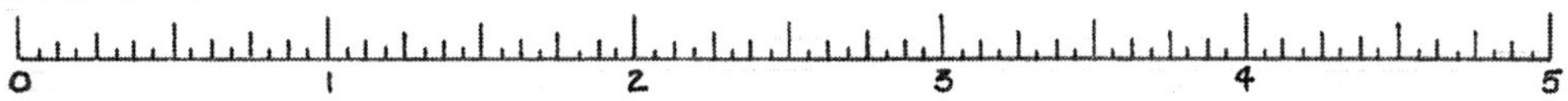

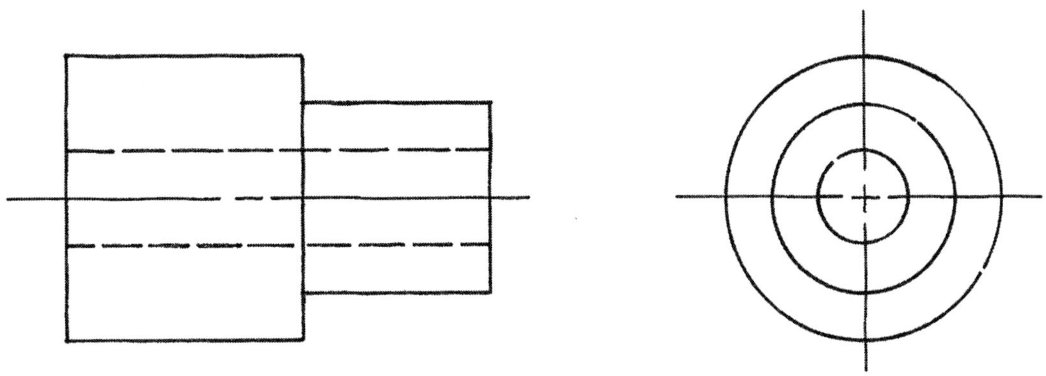

Drawn By:		FREEHAND SKETCHING	
	Scale N/A	TITLE DIMENSIONING	
Date	Sheet of	Drawing Number A	9.1.1

PLACE DIMENSIONS ON ROUNDED CORNERS AND HOLES.

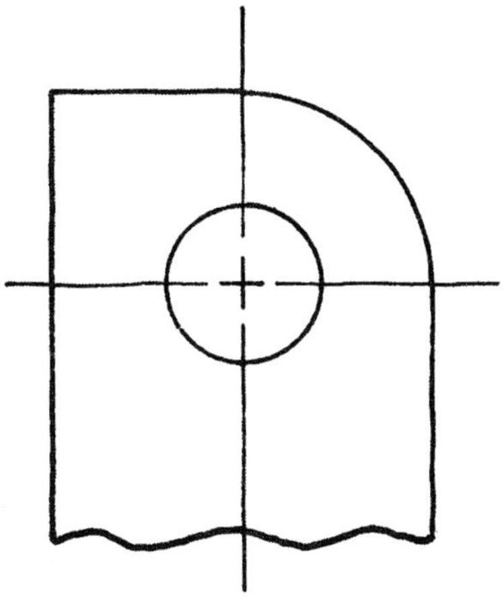

$\frac{3}{4}$ ROUNDED CORNER WITH $\frac{5}{16}$ ϕ HOLE.

PLACE DIMENSIONS ON ROUNDED CORNERS AND HOLES.

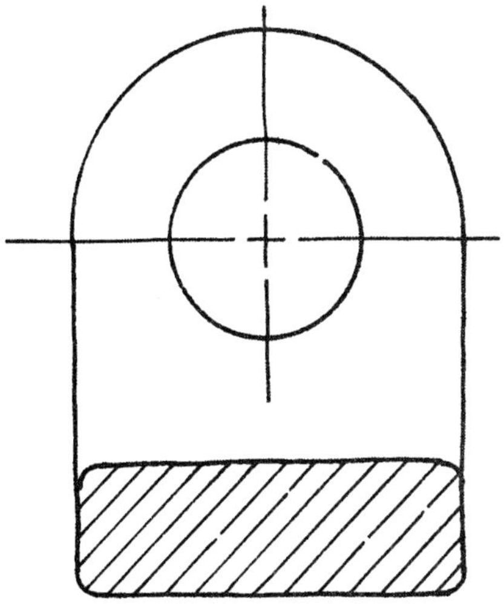

.875 R ROUNDED END WITH A .875 ϕ HOLE.

PLACE DIMENSIONS ON ANGLED CORNERS.

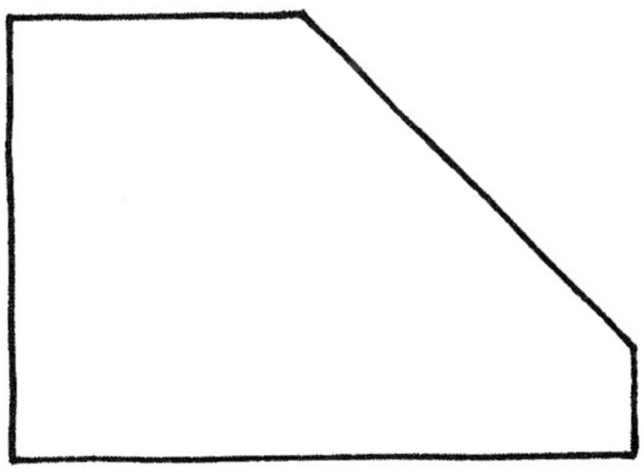

$1\frac{1}{8} \times 45^{\circ}$ ANGLED CORNER BY ANGULAR MEASUREMENT.

PLACE DIMENSIONS ON ANGLED CORNERS.

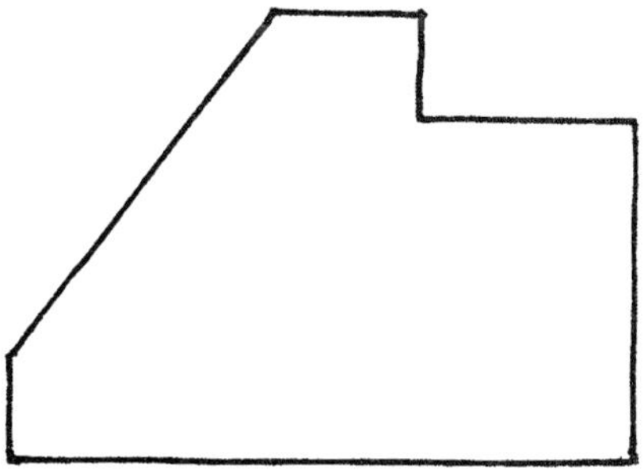

.750 x 1.125 ANGLED CORNER BY LINEAR COORDINATES.

Drawn By:		FREEHAND SKETCHING	
	Scale N/A	TITLE DIMENSIONING	
Date	Sheet of	Drawing Number A	9.1.2

COMPLETE SIZE (S) AND LOCATION (L) DIMENSIONS.
USE (S) AND (L) RATHER THAN NUMERICAL VALUES.

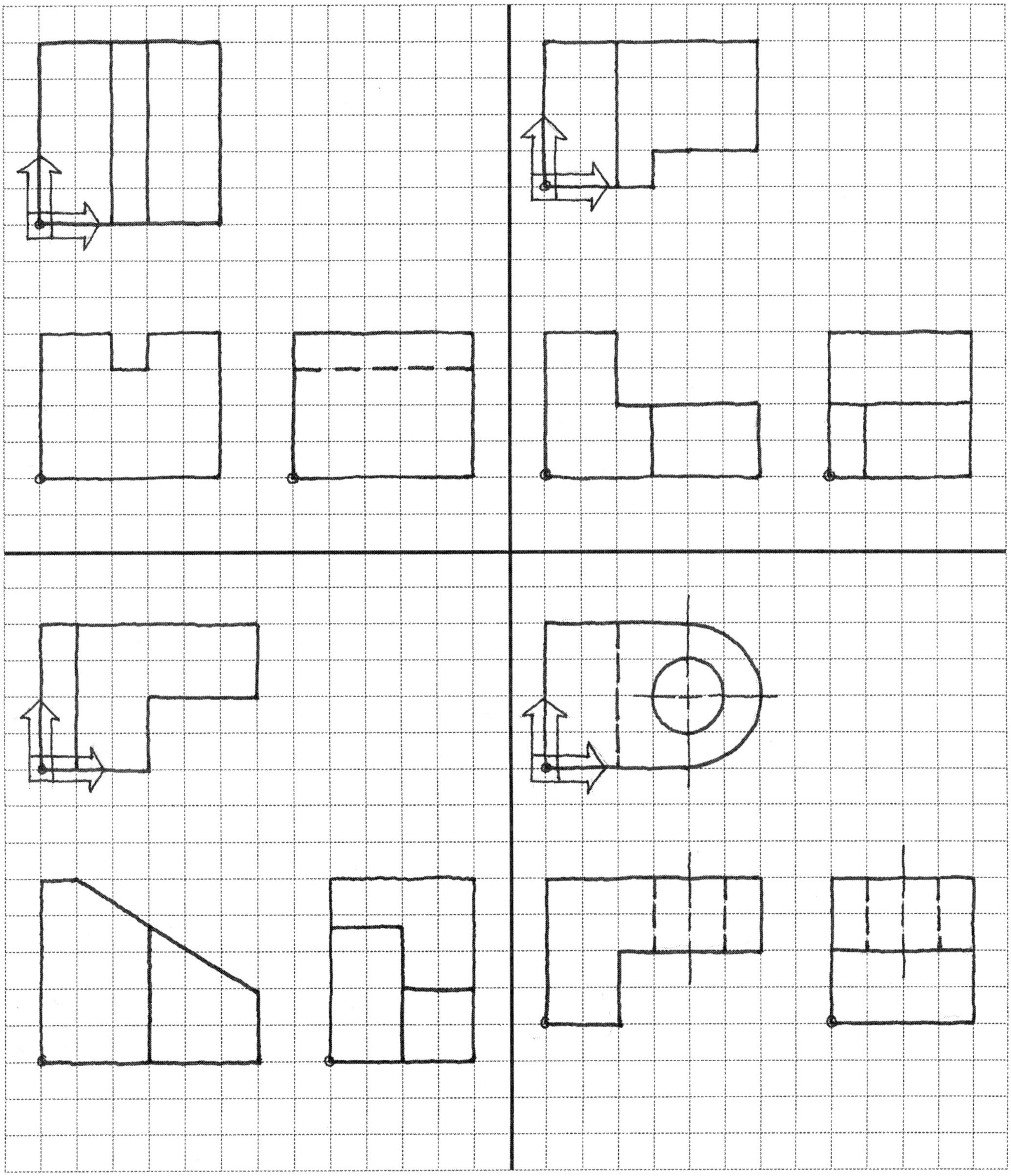

Drawn By:		FREEHAND SKETCHING	
	Scale UNITS	TITLE DIMENSIONING	
Date	Sheet of	Drawing Number A	9.2

PROBLEM A: UNIDIRECTIONAL TO TWO DECIMAL PLACES.
PROBLEM B: ALIGNED - TWO DECIMAL PLACES.

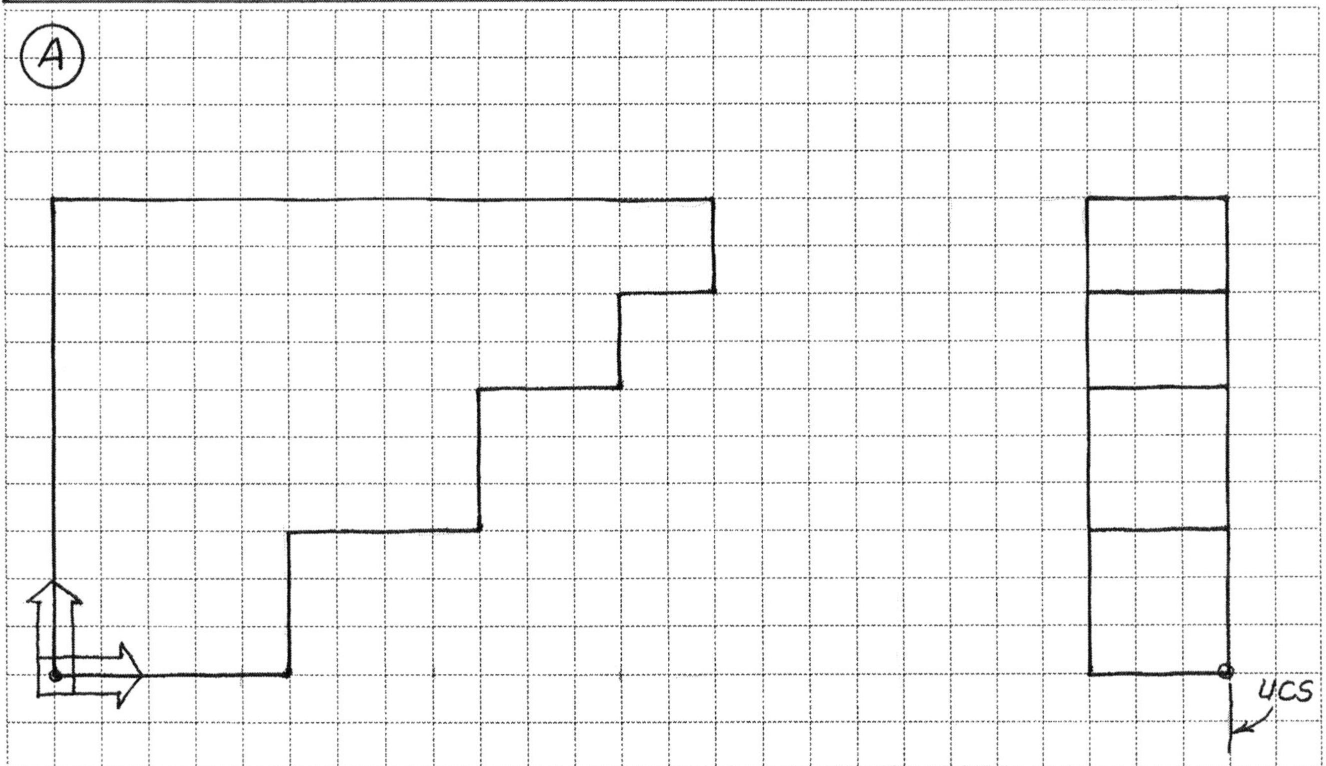

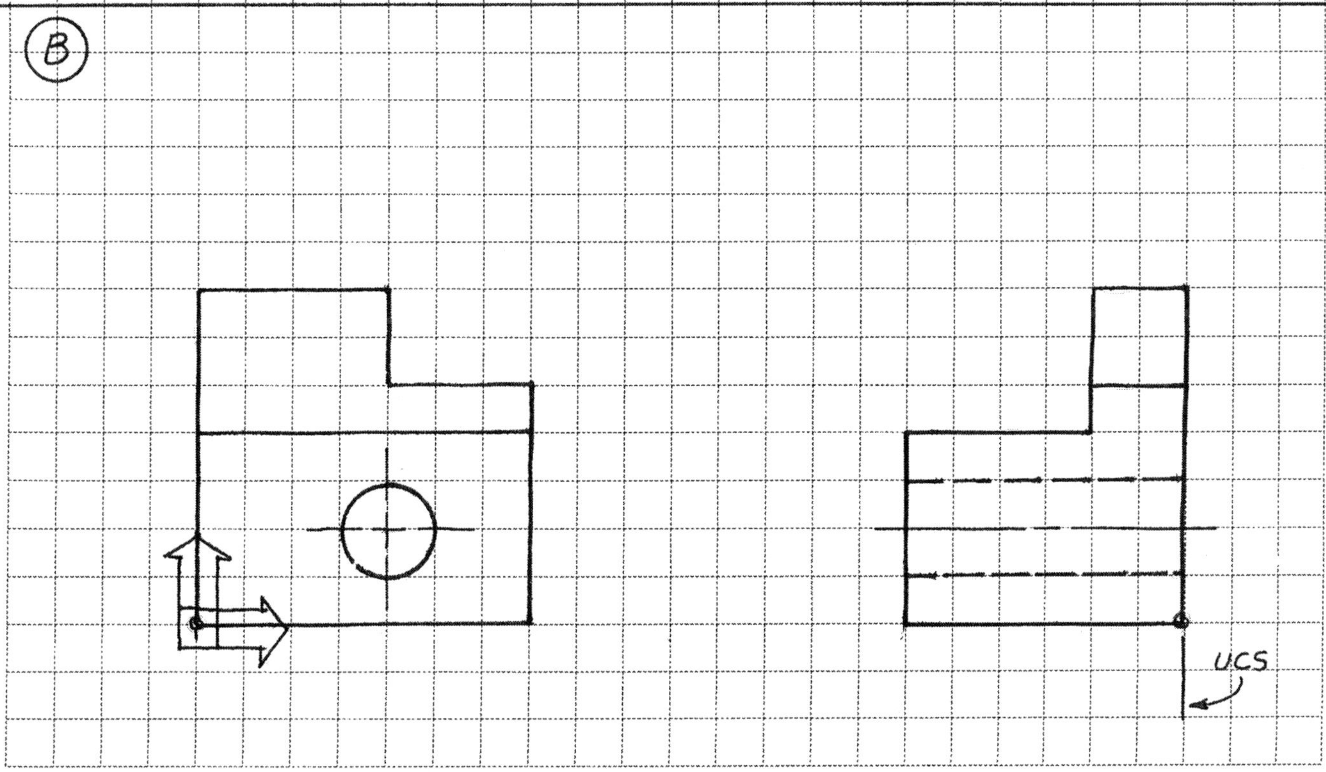

NOTE: .25 GRID

Drawn By:		FREEHAND SKETCHING	
	Scale NOTED	TITLE DIMENSIONING	
Date	Sheet of	Drawing Number A	9.2.1

COMPLETE SIZE (S) AND LOCATION (L) DIMENSIONS.
USE (S) AND (L) RATHER THAN NUMERICAL VALUES.

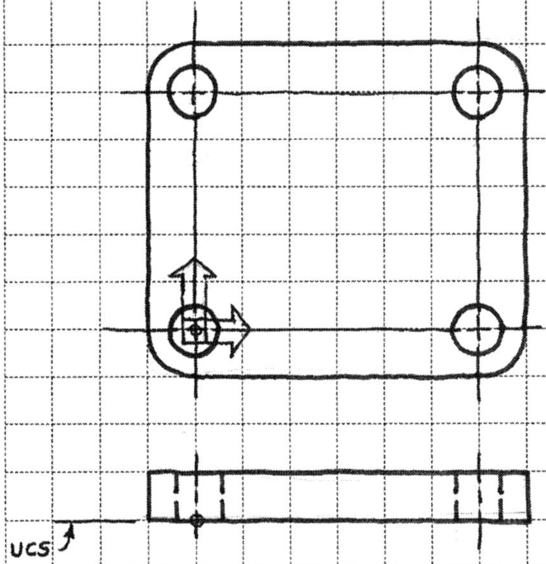

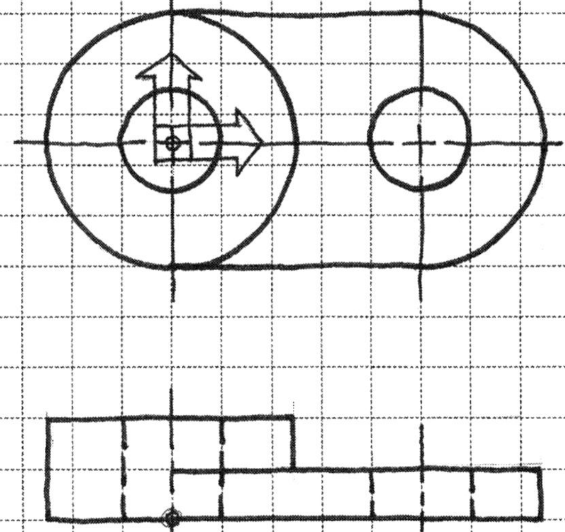

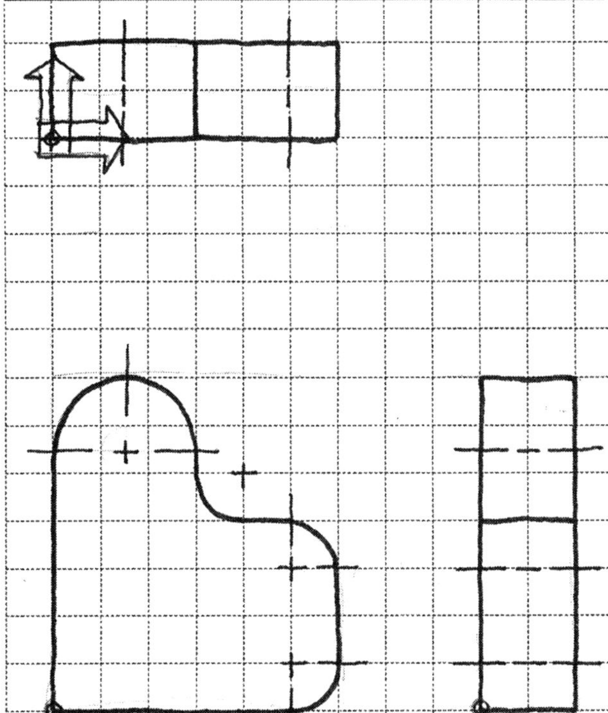

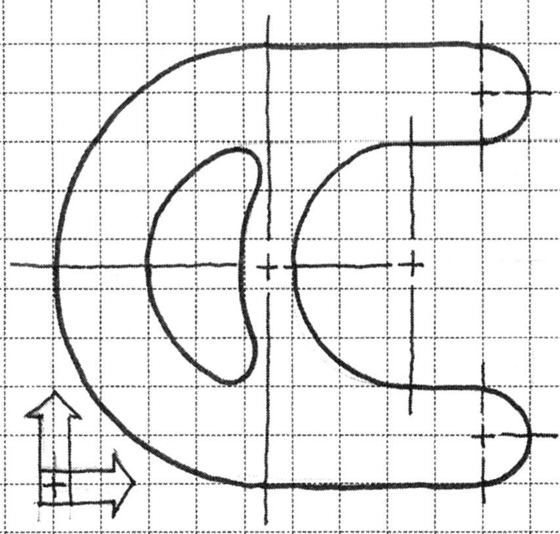

MATERIAL: .075 BRASS

Drawn By:		FREEHAND SKETCHING	
	Scale UNITS	TITLE DIMENSIONING	
Date	Sheet of	Drawing Number A	9.3

PROBLEM A: ADD REQUIRED NOTES & DIMENSIONS. GRID IS .25".
PROBLEM B: ADD REQUIRED NOTES & DIMENSIONS. GRID IS .125."

(A)

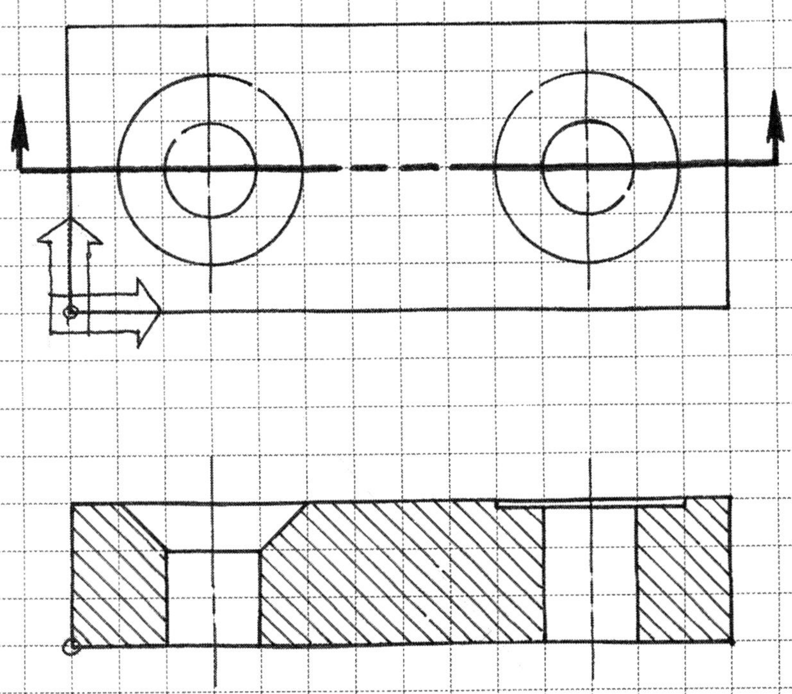

(B)

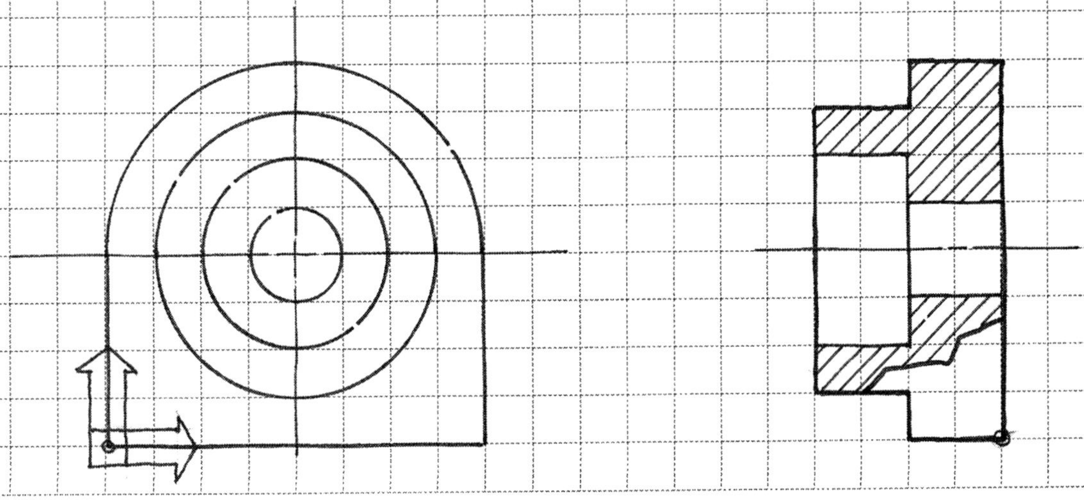

NOTE: BOTH DONE TO TWO DECIMAL PLACES.

Drawn By:		FREEHAND SKETCHING	
	Scale NOTED	TITLE DIMENSIONING	
Date	Sheet of	Drawing Number A	9.3.1

COMPLETELY DIMENSION THE VIEWS. SCALE: FULL

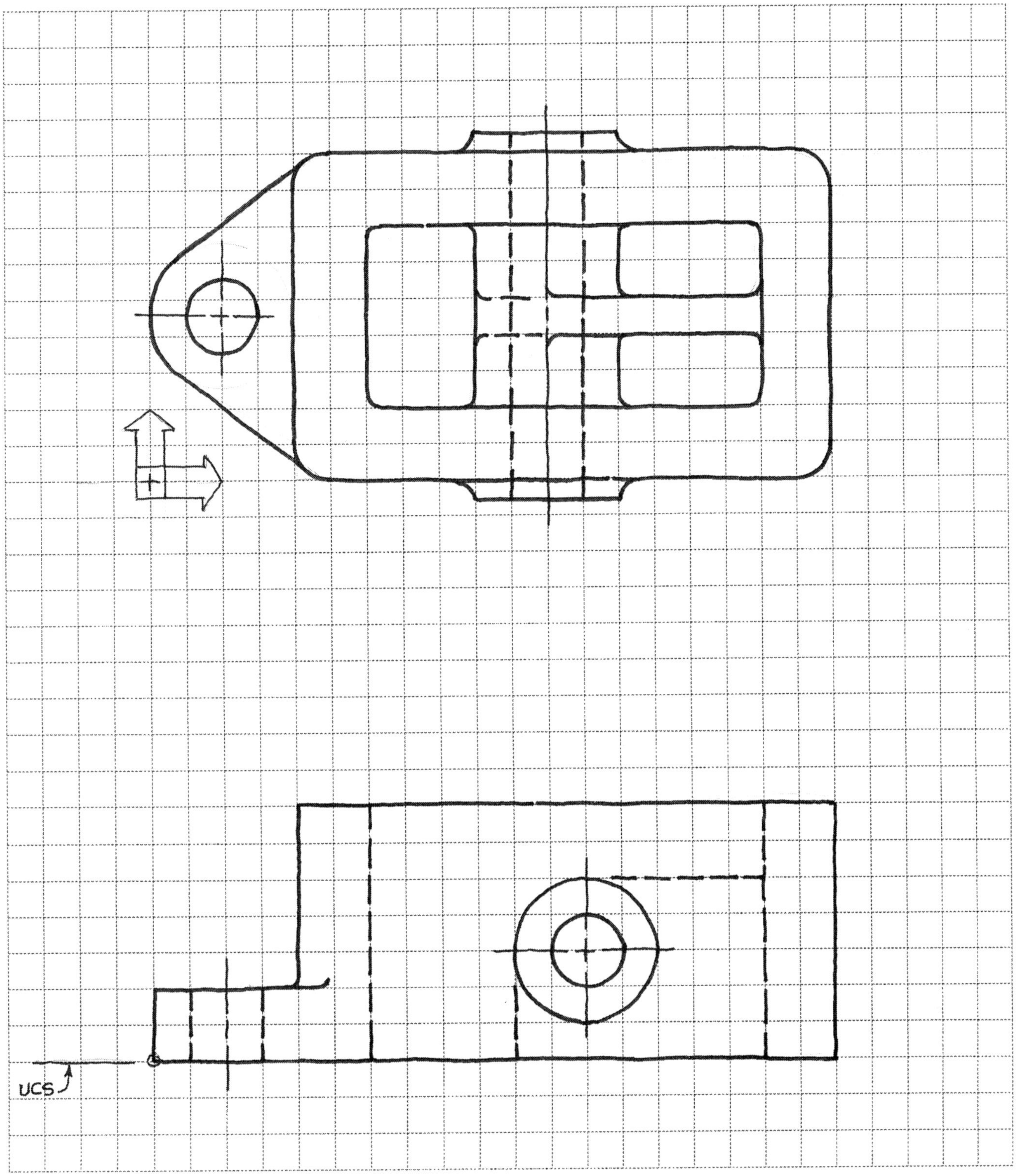

Drawn By:		FREEHAND SKETCHING	
	Scale 1:1	TITLE DIMENSIONING	
Date	Sheet of	Drawing Number A	9.4

PROBLEM A: NOTES & DIMENSIONS TO TWO DECIMAL PLACES. GRID=.125"
PROBLEM B: NOTES & DIMENSIONS TO TWO DECIMAL PLACES. GRID=.25"

(A)

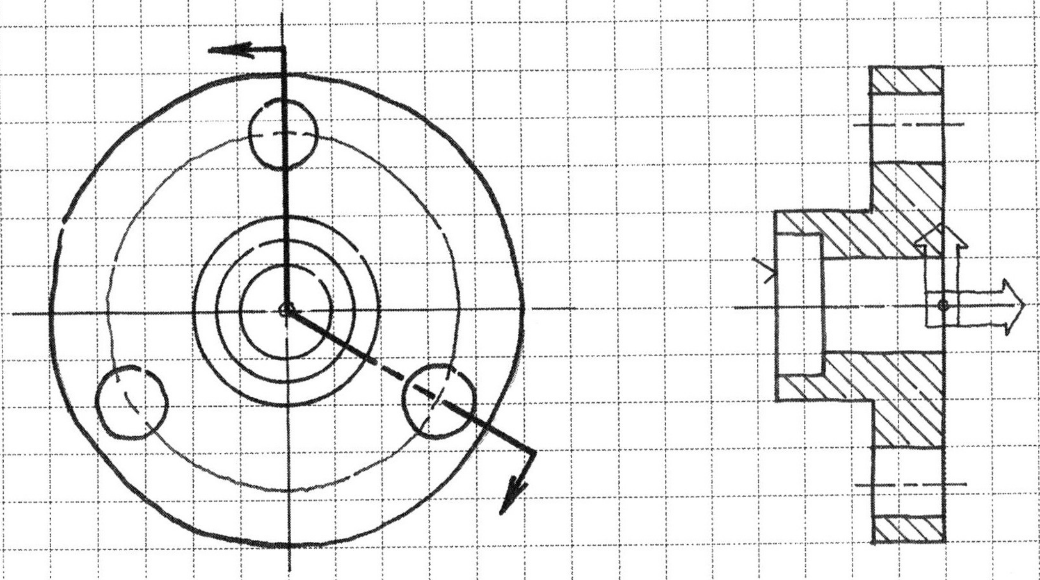

(B)

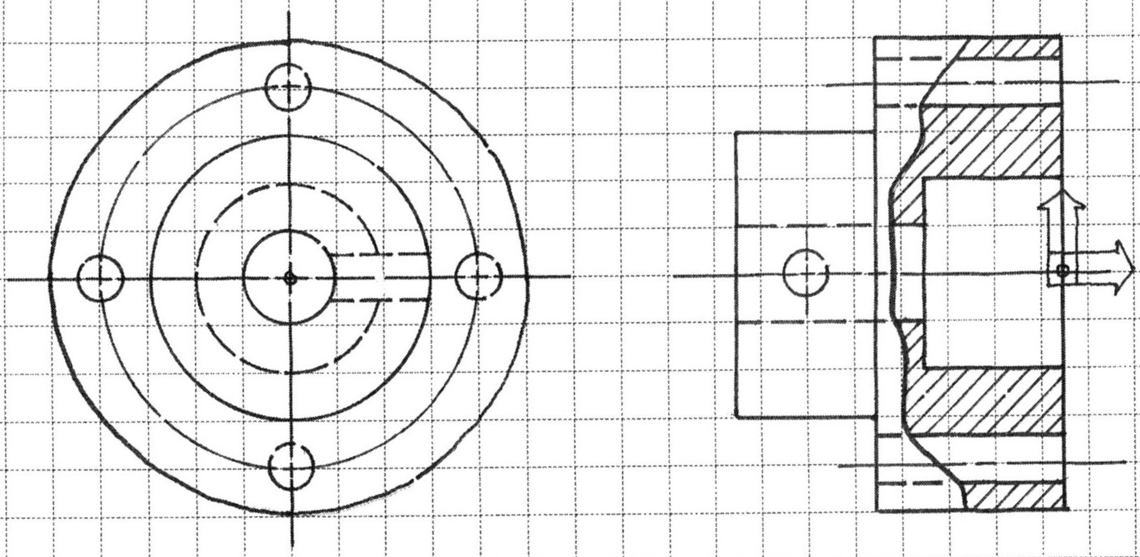

Drawn By:		FREEHAND SKETCHING	
	Scale NOTED	TITLE DIMENSIONING	
Date	Sheet of	Drawing Number A	9.4.1

DETERMINE SCALE AND FULLY DIMENSION THE VIEWS.

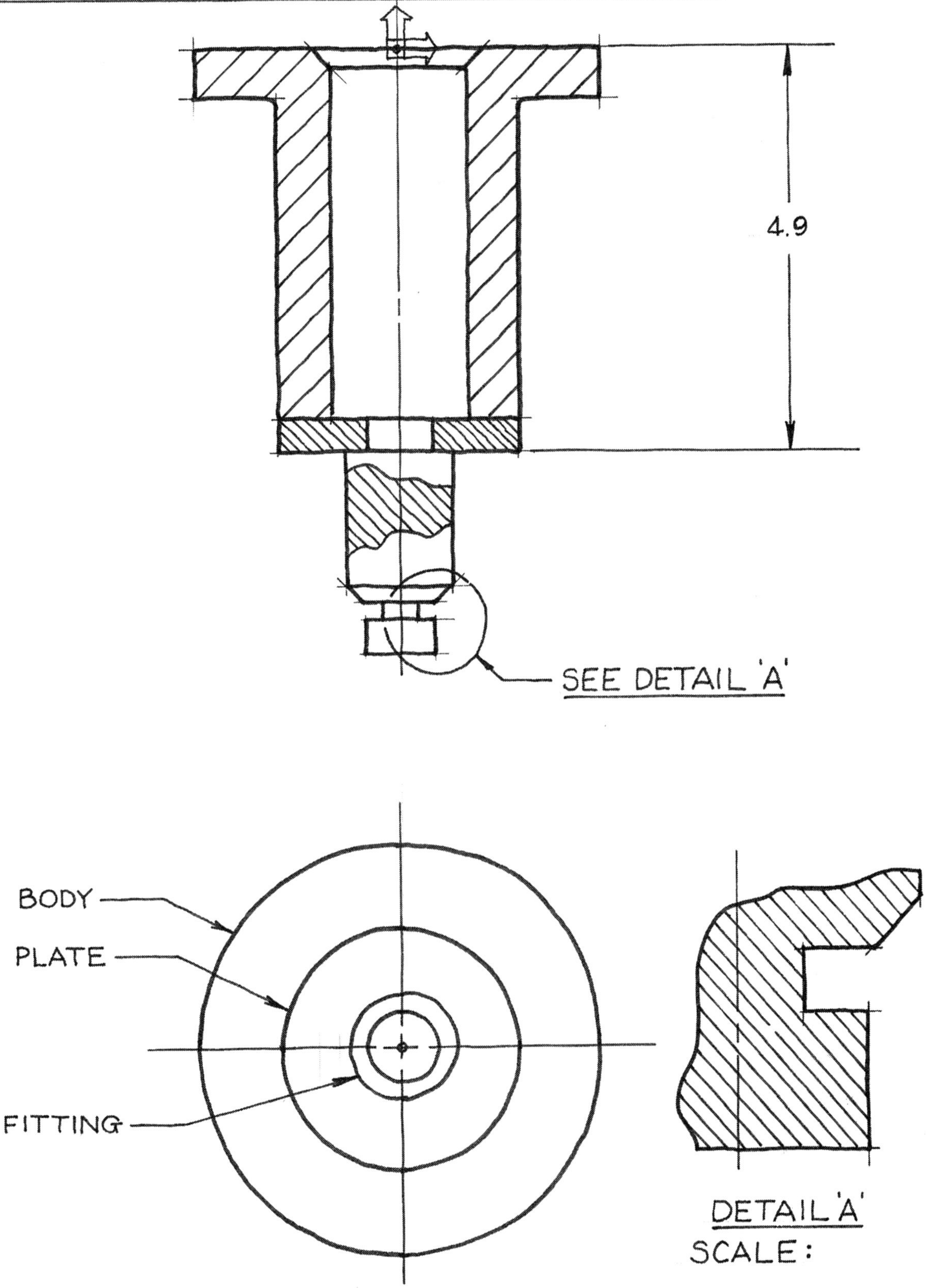

Drawn By:		FREEHAND SKETCHING	
	Scale	TITLE DETERMINE SCALE	
Date	Sheet of	Drawing Number A	9.5

DEVELOP A SCALE FOR THIS SECTIONAL VIEW.
RECORD DIMENSIONS IN TABULAR FORM.

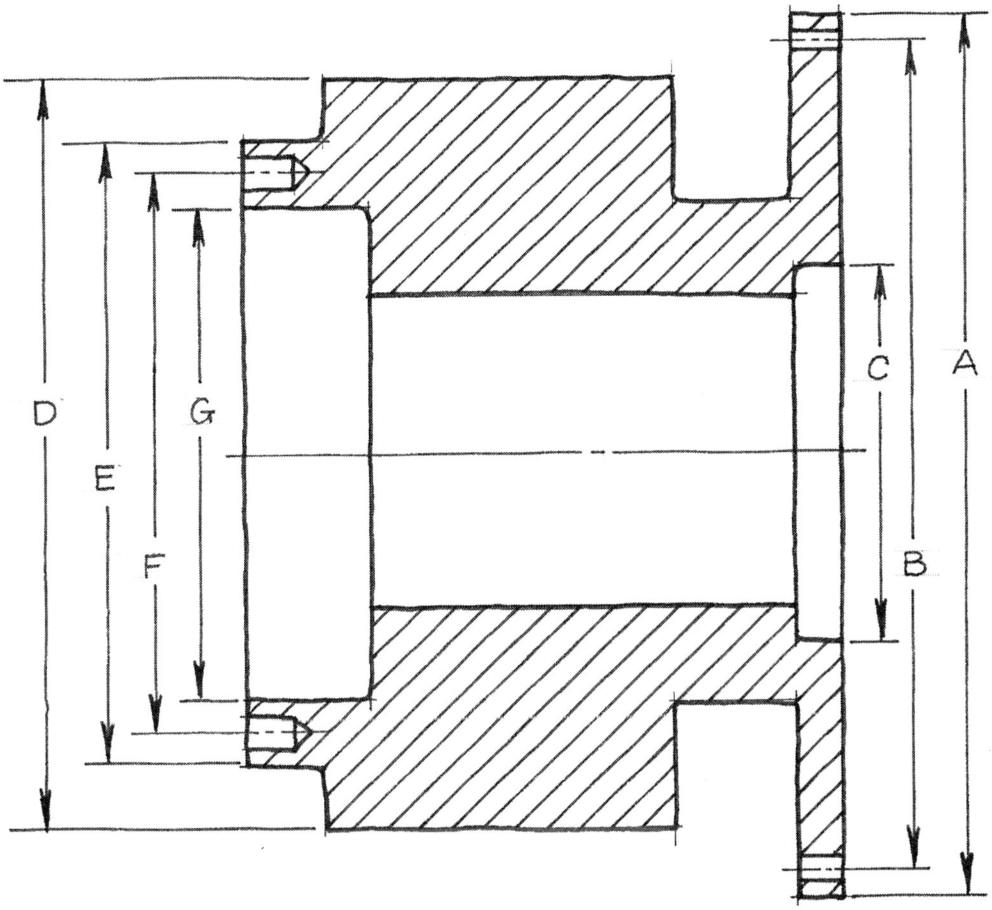

TABULAR DIMENSIONS						
A	B	C	D	E	F	G
8.6						

Drawn By:		FREEHAND SKETCHING	
	Scale 1:1	TITLE DIMENSIONING	
Date	Sheet of	Drawing Number A	9.6

RECORD VALUES IN DATUM DIMENSIONING FORMAT.

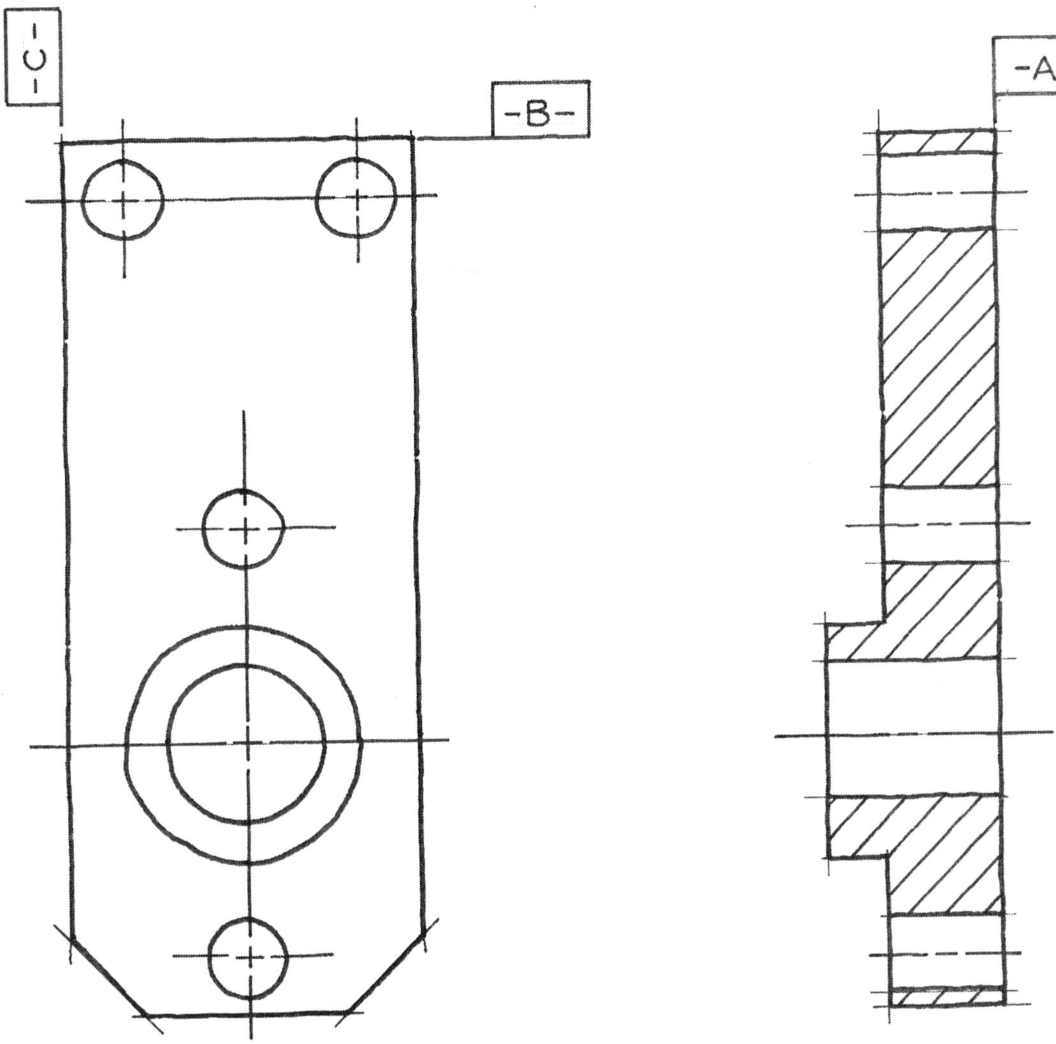

TERMINAL TOP

Drawn By:		FREEHAND SKETCHING	
	Scale 1:1	TITLE DIMENSIONING	
Date	Sheet of	Drawing Number A	9.7

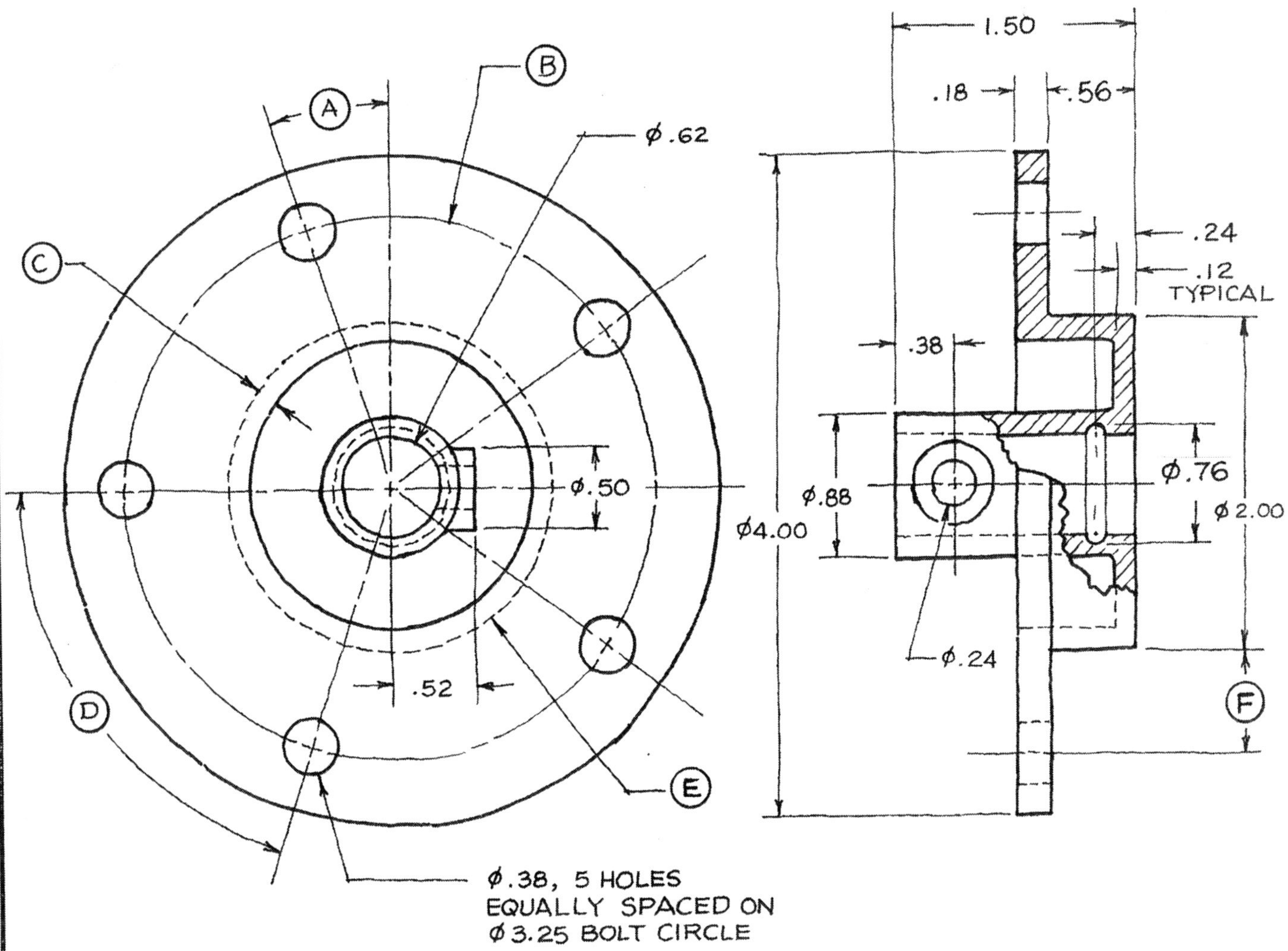

QUESTIONS:

1. What type of sectional view is shown?
2. What is angle (A)?
3. What is the thickness at (C)?
4. How thick is the largest flange?
5. What is diameter (B)?
6. What is angle (D)?
7. What is diameter (E)?
8. What is dimension (F)?

ANSWERS:

1. __________
2. __________
3. __________
4. __________
5. __________
6. __________
7. __________
8. __________

Drawn By:		FREEHAND SKETCHING	
	Scale NTS	TITLE READING EXERCISE	
Date	Sheet of	Drawing Number A	9.8

10 Fabrication Sketching

Coordinated with VMGED Chapter 9

Why Sketch for Fabrication?

When you plan model geometry by sketching, you are generally concerned with size, shape, and location. But these models will probably be fabricated—assembled from standard or altered commercial shapes, cast, formed or molded, machined from solid material, or created by a combination of methods. Your sketches may need to reflect and communicate fabrication procedures. For these reasons, you will want to consider potential fabrication methods as you sketch.

Figure 10-1 displays sketches of a standard fastener, a commercial assembly, and a fabricated part. Notice that the sketches contain realistic details. These details go a long way toward representing how the products are fabricated.

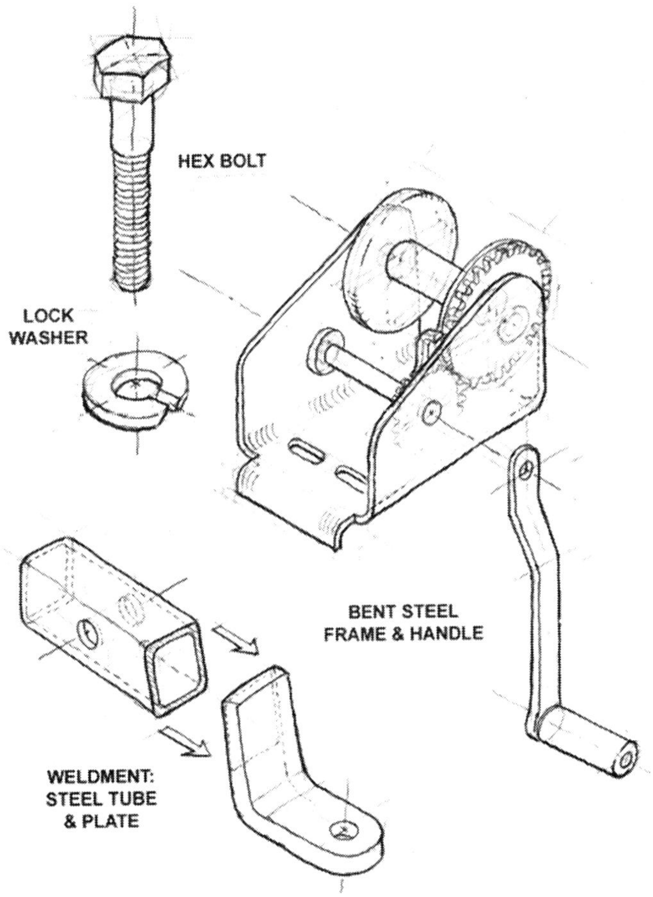

Fig. 10-1. Fabricated commercial products.

Standard Commercial Shapes

Standard commercial shapes may be thought of as geometric primitives. Many of these shapes are formed by extrusion, a process you can easily replicate through sketching. In Fig. 10-2, a C-channel cross-section has been sketched on a frontal construction plane. That cross-section has been extruded pictorially in a perpendicular direction and the corresponding corners of the cross-sections connected. Finally, sketch lines have been darkened for visibility.

Most commercial constructions are made by cutting and fastening standard commercial shapes—I-beams, angle beams, channels, and flat plates. Even the houses we live in are fabricated from studs, rafters, trusses, and plywood sheets that begin as standard-sized commercial shapes. Note how the C-channel shape has been used in the trailer assembly.

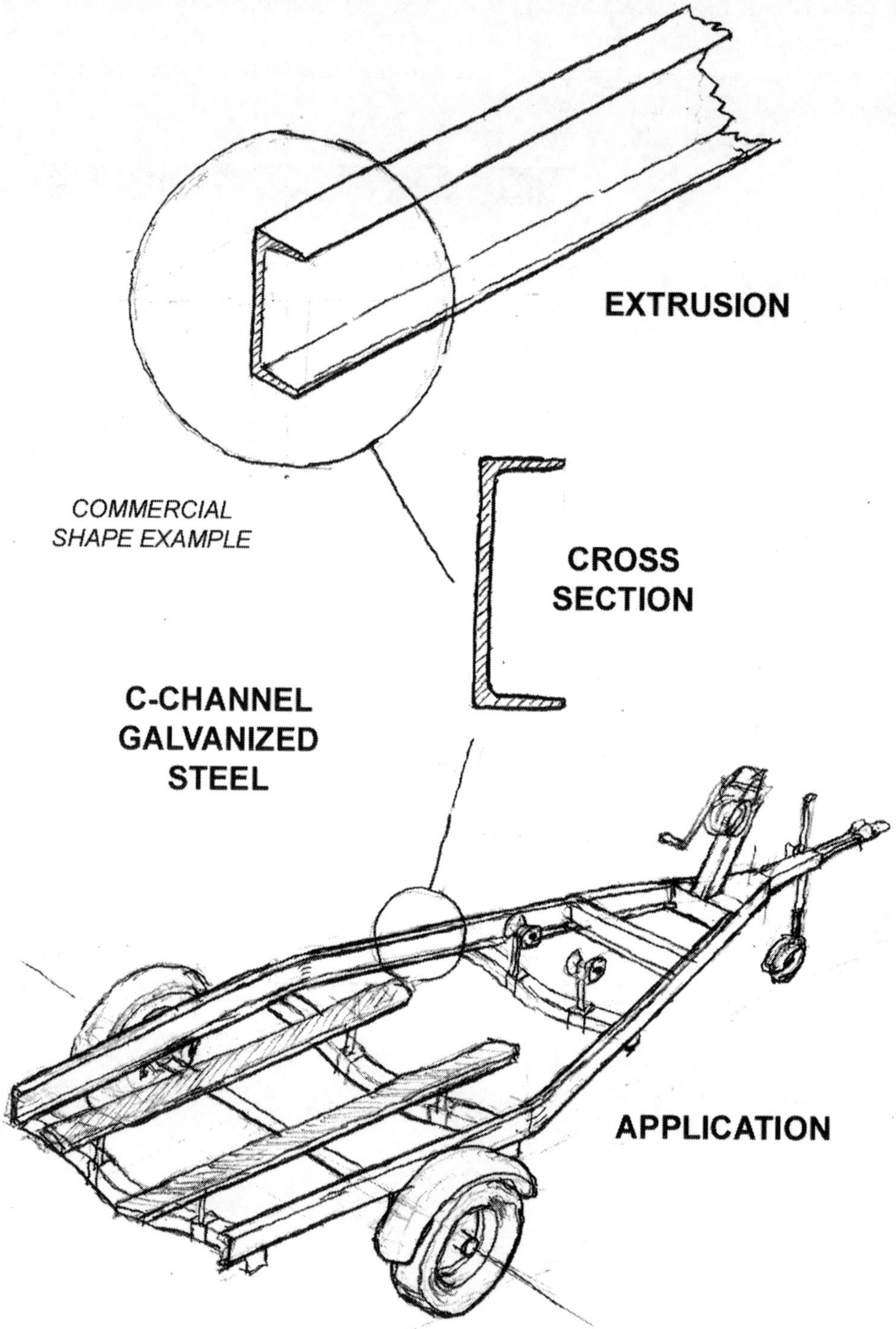

Fig. 10-2. Extruding a common commercial shape and a product making use of that component.

Folded Flat Patterns

Many fabricated products are made from sheets of material that have been cut, folded, and then fastened by a variety of means. Common examples are air conditioning and heating ductwork and packaging of all types. The duct in Fig. 10-3 begins as a sheet of common sheet stock. The sides of the duct are drawn on the sheet, along with tabs for securing the final shape. Dashed lines are used to communicate where folds are located. The tabs are first folded 90°, then, the planes. Finally, rivets are shown securing the tabs to the sides. A person should be able to follow this sketch and understand how the duct is made.

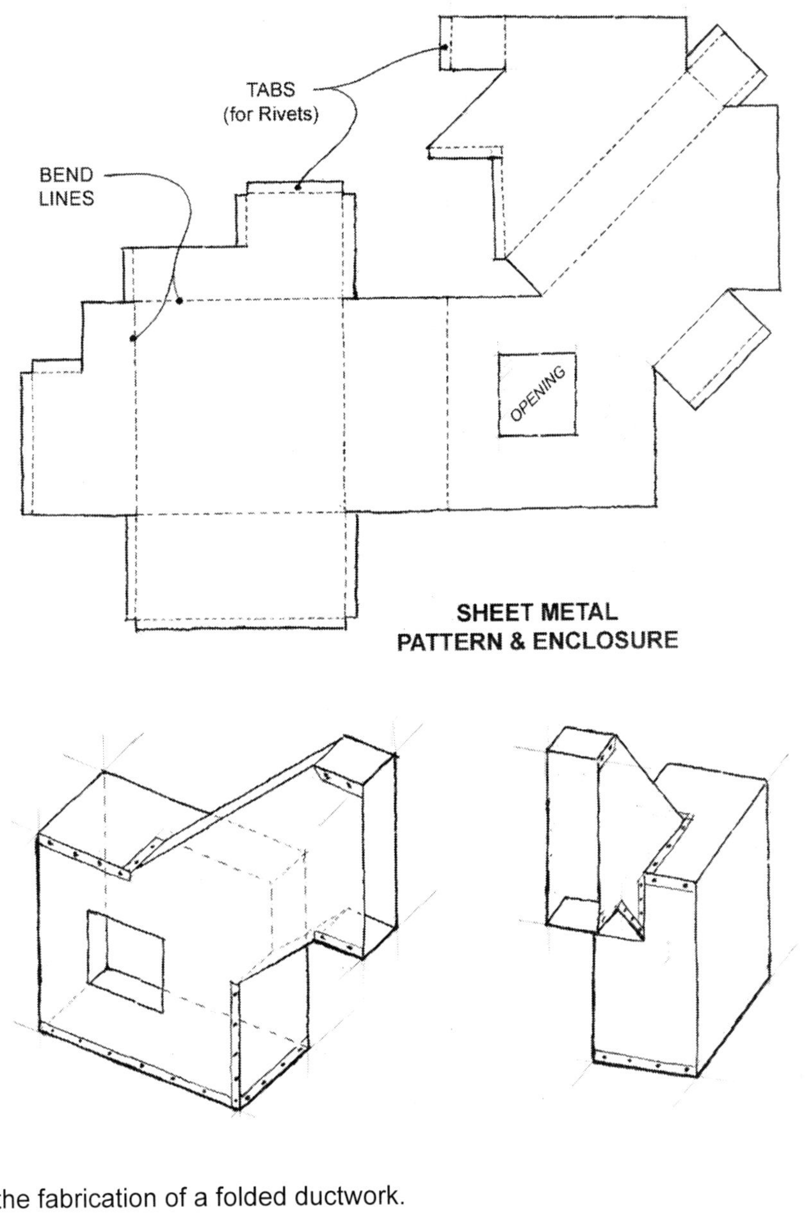

Fig. 10-3. Sketching the fabrication of a folded ductwork.

Fillets and Rounds

It is unwise to create sharp corners on just about any product. Not only is it difficult to create sharp corners, the small amount of material at the tip may be brittle and break, or may catch on other parts. A sharp corner that has been angled is called a **chamfer**. An exterior corner smoothed with a radius is called a **round**. An interior corner smoothed with a radius is called a **fillet**.

We show these altered corners in sketches to better communicate function and the type of fabrication process necessary to make the part.

Figure 10-4 shows a part where chamfers, fillets, and rounds have been applied. Fillets and rounds are circular radii tangent to the two surfaces they span. These tangents are important because they may aid in perceiving that a corner is, in fact, filleted or rounded. Chamfer lines should be thin and sharp. Fillets and rounds are easily visualized on a box because of rounded exterior profiles. On a cylinder, however, it is the addition of hatched sketch lines at the tangents that helps communicate the round.

Observe the differences between the prism and cylinder intersection (marked with circles) in Fig. 10-4. Without the hatch lines the intersection cannot be easily visualized. With the hatch lines, the intersection is clear.

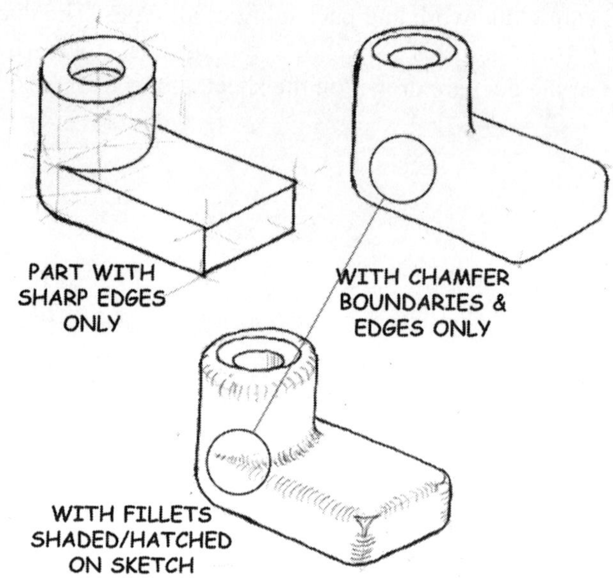

Fig. 10-4. Chamfers, fillets, rounds, and tangent lines.

Drilling Holes

In pictorial sketches, drilled holes are represented by ellipses. Refer to Chapter 3 for methods of drawing curves and ellipses, especially in the way that a hole's diameter can only be measured along pictorial axes. A drilled hole may pass through an object, or be drilled only part way. If the material is thin, the exit of the hole (Fig. 10-5) can easily be seen. If the material is thick, you may want to sketch dashed lines to show its depth. In most cases, the conical end of the drilled hole is omitted. The length of the hole's limiting element can then be compared to the material thickness to determine if the hole is drilled completely through. Inspect Fig. 10-5 and make decisions as to the depth of the drilled holes. Is there a hole whose depth is impossible to determine?

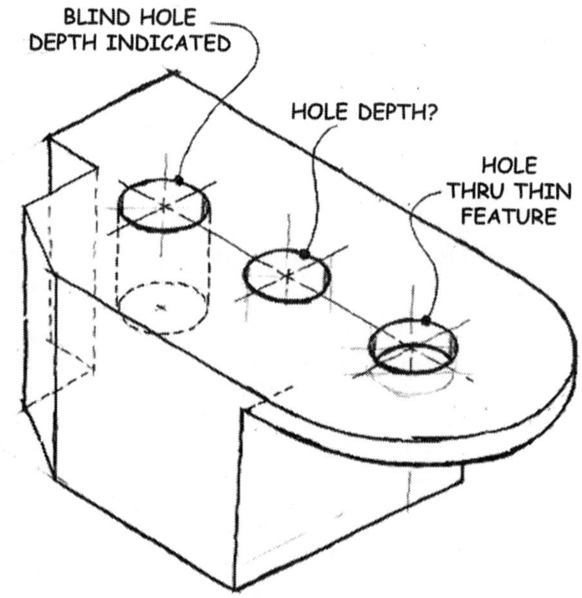

Fig. 10-5. Drilling holes.

Turned Shapes

Turning produces geometry symmetrical about an axis. Because the path of turned geometry is circular about the axis, it's important to be able to sketch accurate ellipses. Figure 10-6 shows a profile and an axis, ready for turning. Notice how the placing of ellipses at critical points along the profile determines the outer limits of the turned shape. The center of each ellipse is on the axis; the centerlines define the radius of each circle. The outside profile of the turned shape is sketched either as straight lines (between two ellipses of the same diameter) or as a smooth curve, tangent to the ellipses.

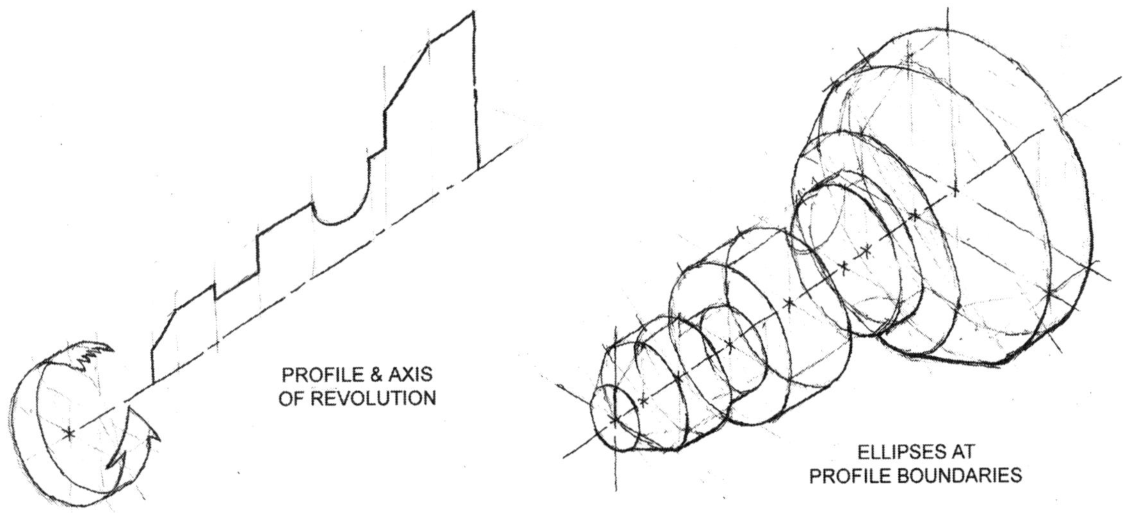

Fig. 10-6. Sketching a turned shape.

Milled Shapes

A milling machine uses a specially shaped tool to remove material by cutting. The shape of the tool determines the shape of the removed material. For example, say you want to remove the dovetail-shaped groove in Fig. 10-7. As shown, a tool could be designed to create the groove in one pass. Or, a standard tool could cut the first part of the groove, and another standard tool could cut the remainder. This can be thought of as a "negative extrusion," and is sketched in a fashion similar to Fig. 10-2.

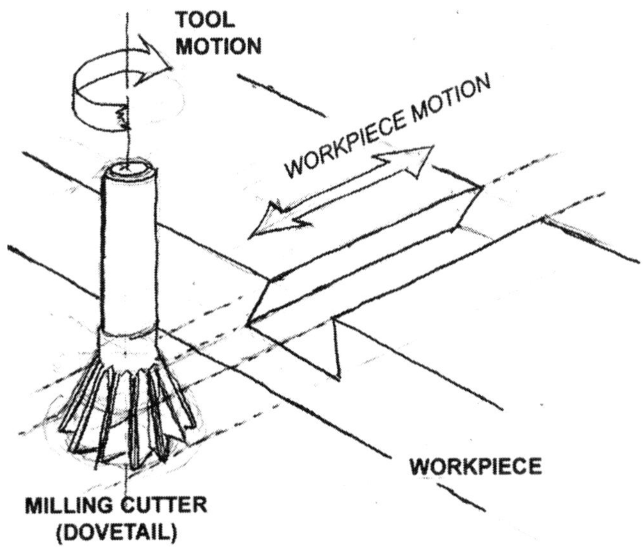

Fig. 10-7. Sketching a milled shape.

Welding and Casting

A single part can be fabricated by welding together a number of smaller parts. Showing a welded bead in a sketch (Fig. 10-8) is accomplished with irregular line work. You can even put in a few irregular lines to portray the rounded nature of the welded bead.

For higher volume production, a mold or die is made to form the product in one piece. In a sketch, the welded beads are replaced with fillets and rounds, added as necessary on interior and exterior corners. Still, holes and precision surfaces must be machined. The same part formed previously by welding multiple pieces is also shown in Fig. 10-8 as a single cast part. Note the use of fillets and rounds to minimize sharp interior and exterior corners, and the lines in the hole that give the impression of a machined surface.

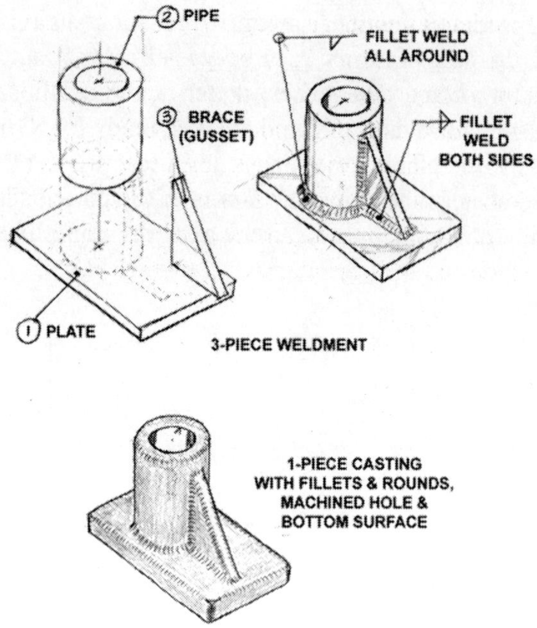

Fig. 10-8. A part made by welding and then casting.

CAD Example—2D

Two-dimensional parts such as gaskets, shims, and plates are created by a number of cutting methods including sawing, laser and plasma cutting, and cutting by high-speed water jets. These cutting tools are directed by a programming language called G-code. This code tells the machine which tool to use, and how to move it. The first step is to know the precise location of each feature. Figure 10-9 establishes X- and Y-datums and the location of each feature relative to those datums.

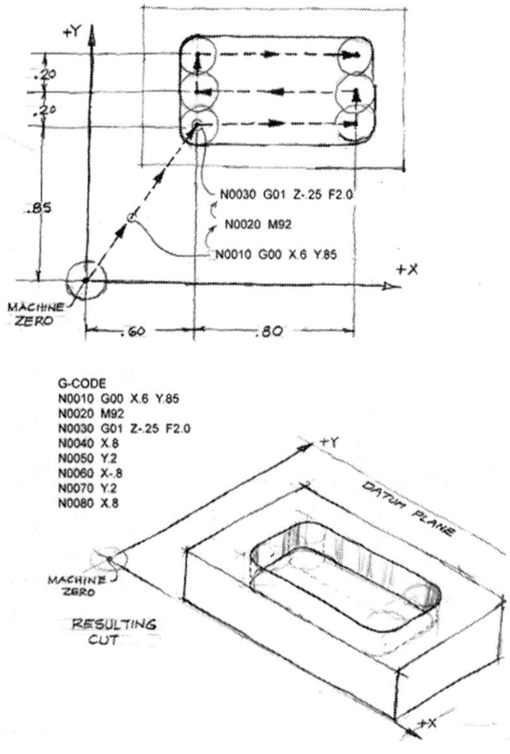

Fig. 10-9. A sketch locating X-Y positions relative to datums.

CAD Example—3D

Three-dimensional parts also require datum planes to precisely control fabrication. Each of these planes originally contains two of the three machine axes (corresponding to the three world axes). If a part has angular or oblique features, these axes can be repositioned relative to a local axis direction. Inspect the part shown in Fig. 10-10. Three separate machining axes systems have been identified in order to face and drill the angular mounting plates.

Sketching Exercises

The sketching exercises on the following pages will help you apply your knowledge of fabrication processes. Sheet 10.1 is an example of how drilling and milling can be used to create a finished part (called the workpiece).

Sheet 10.5 shows how a flat pattern can be developed that results in a folded shape. You may wish to transfer your sketch onto heavier paper and actually cut the pattern out. Make sure you include the tabs and glue or tape the model together for rigidity.

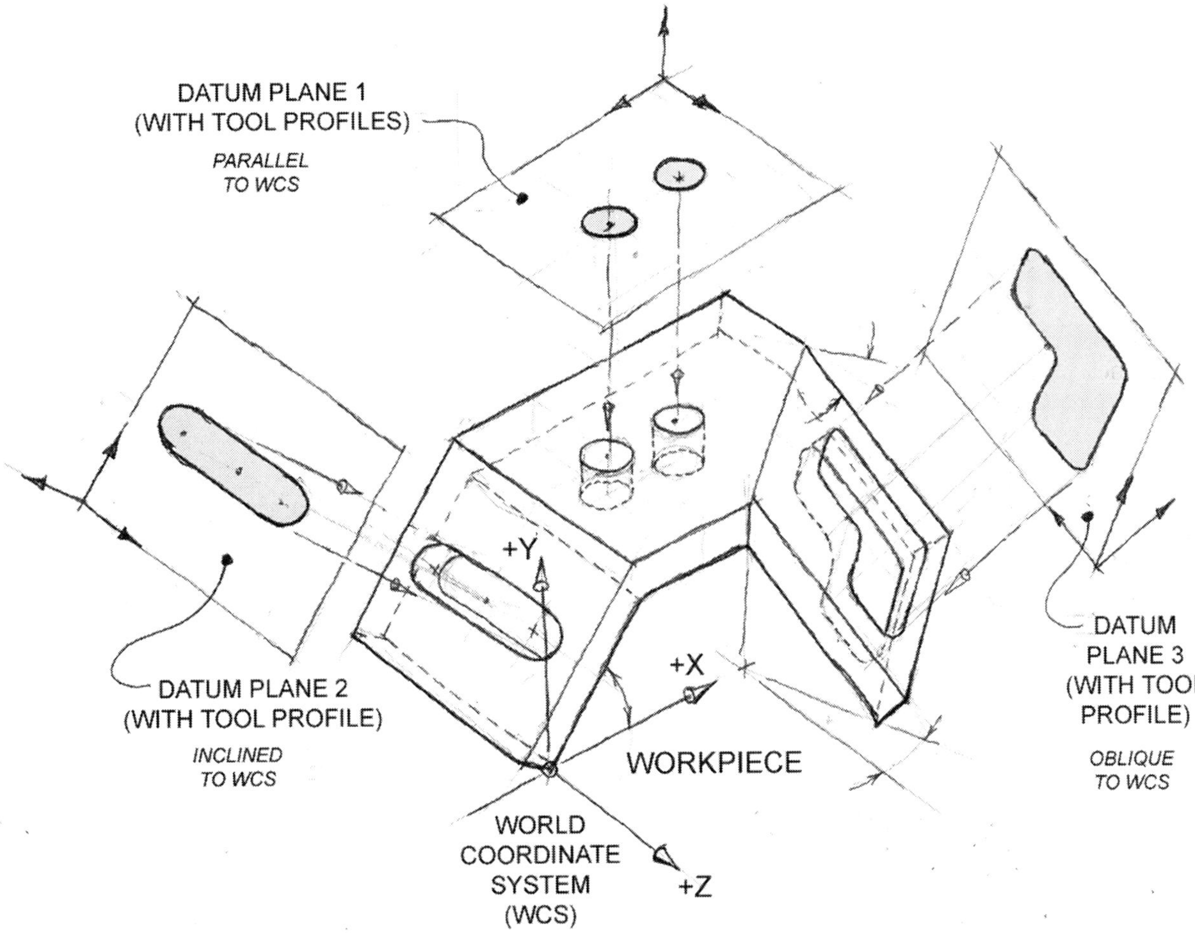

Fig. 10-10. 3D fabrication calls for multiple machining axes.

SKETCHING FABRICATION-CAD MODELING PROCESSES: Shown below is the sketch of a workpiece and the results of what the workpiece looks like after performing a series of 3D cut or drill machining operations. In solid modeling CAD these operations are represented by datum planes and 2D "tool" profiles that sweep through and "cut out" the resulting shapes as shown in the sketch of the finished workpiece.

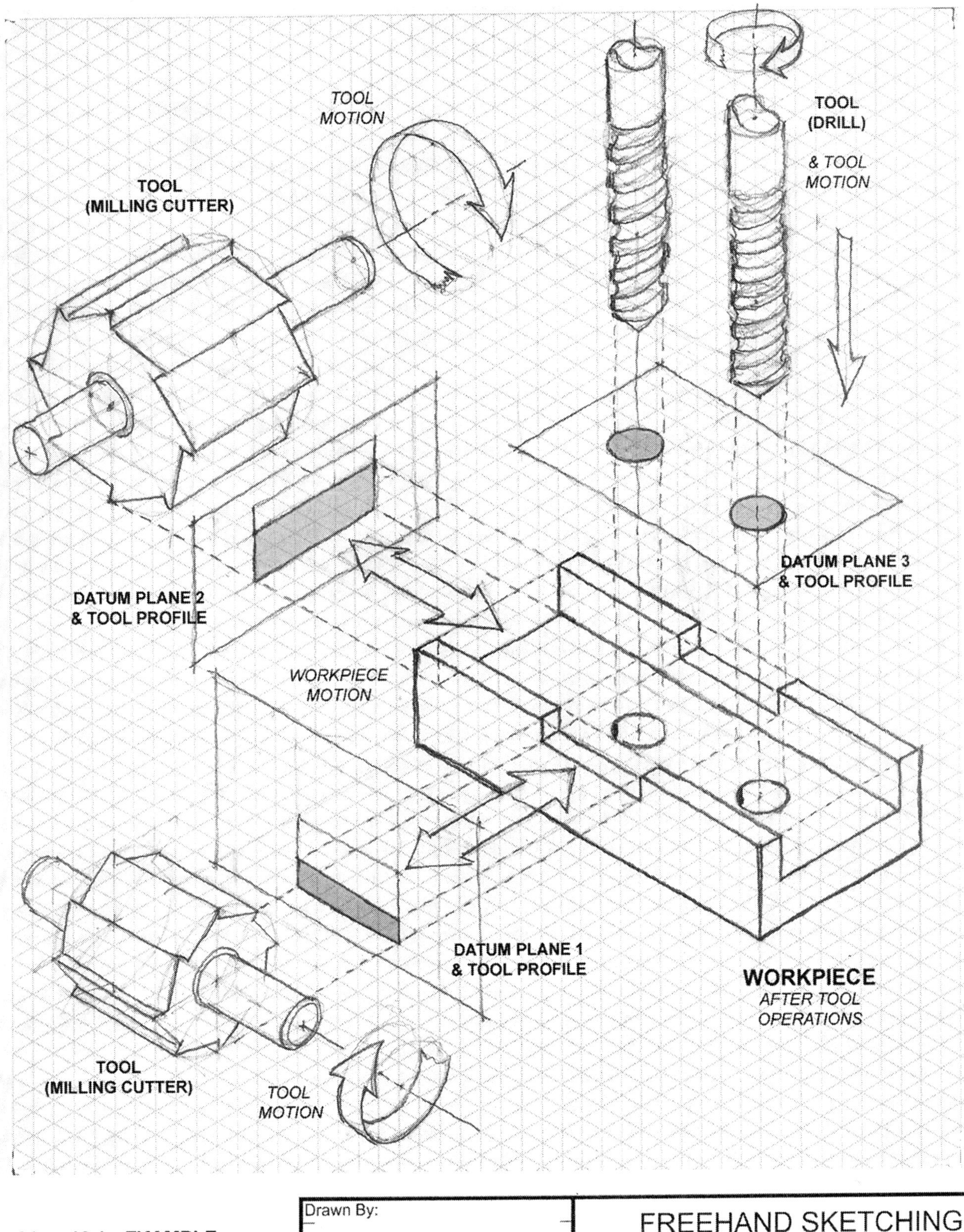

Problem 10.1. ***EXAMPLE***

Drawn By:		FREEHAND SKETCHING	
	Scale	TITLE FABRICATION	
Date	Sheet	Drawing Number A	10.1

SKETCHING FABRICATION-CAD MODELING PROCESSES: Given the extrusion profile and the tool profiles shown below, sketch an isometric of the finished results of the machining CAD modeling operations on the workpiece in the space bounded between corners A - B and C - D. If assigned, include hidden lines in your isometric sketch.

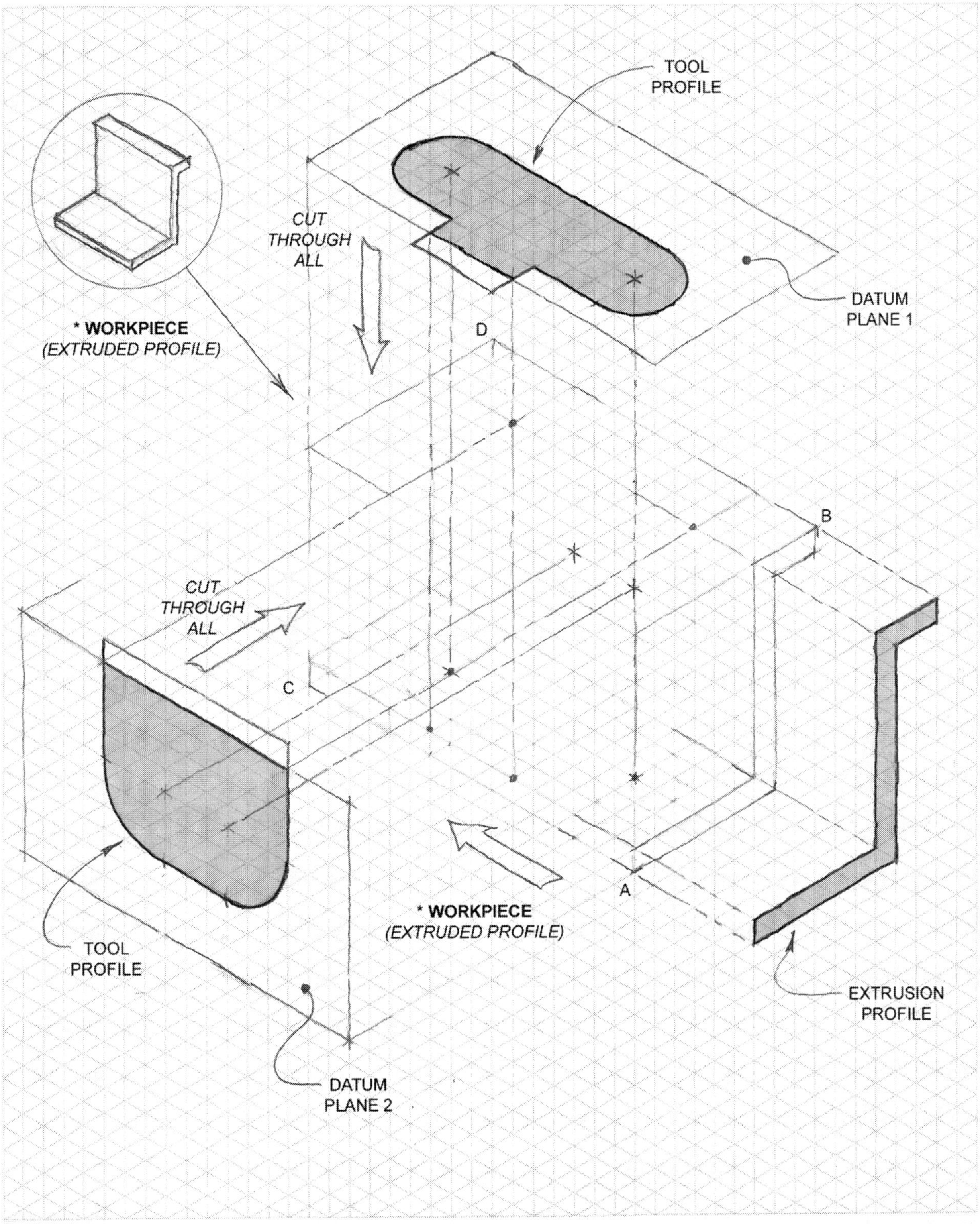

Problem 10.2

Drawn By:		FREEHAND SKETCHING	
	Scale	TITLE FABRICATION	
Date	Sheet	Drawing Number A	10.2

SKETCHING FABRICATION-CAD MODELING PROCESSES: Given the extrusion profile and the tool profiles shown below, sketch an isometric of the finished results of the machining CAD modeling operations on the workpiece in the space bounded by corners A thru G. If assigned, include hidden lines in your isometric sketch.

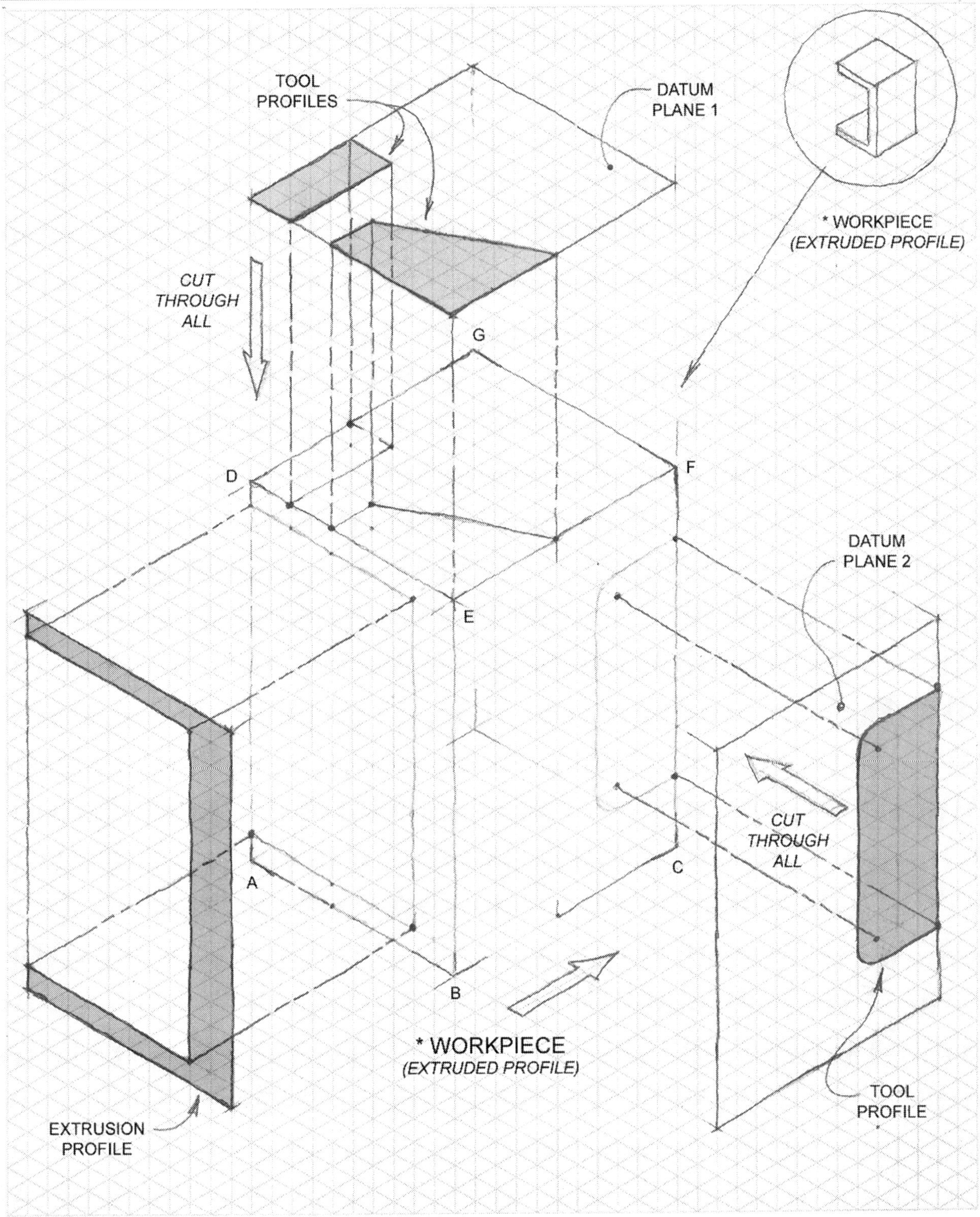

Problem 10.3

Drawn By:		FREEHAND SKETCHING	
	Scale	TITLE FABRICATION	
Date	Sheet	Drawing Number A	10.3

SKETCHING FABRICATION-CAD MODELING PROCESSES: Given the extrusion profile and the tool profiles shown below, sketch an isometric of the finished results of the machining CAD modeling operations on the workpiece in the space bounded between corners A - B and C - D. If assigned, include hidden lines in your isometric sketch.

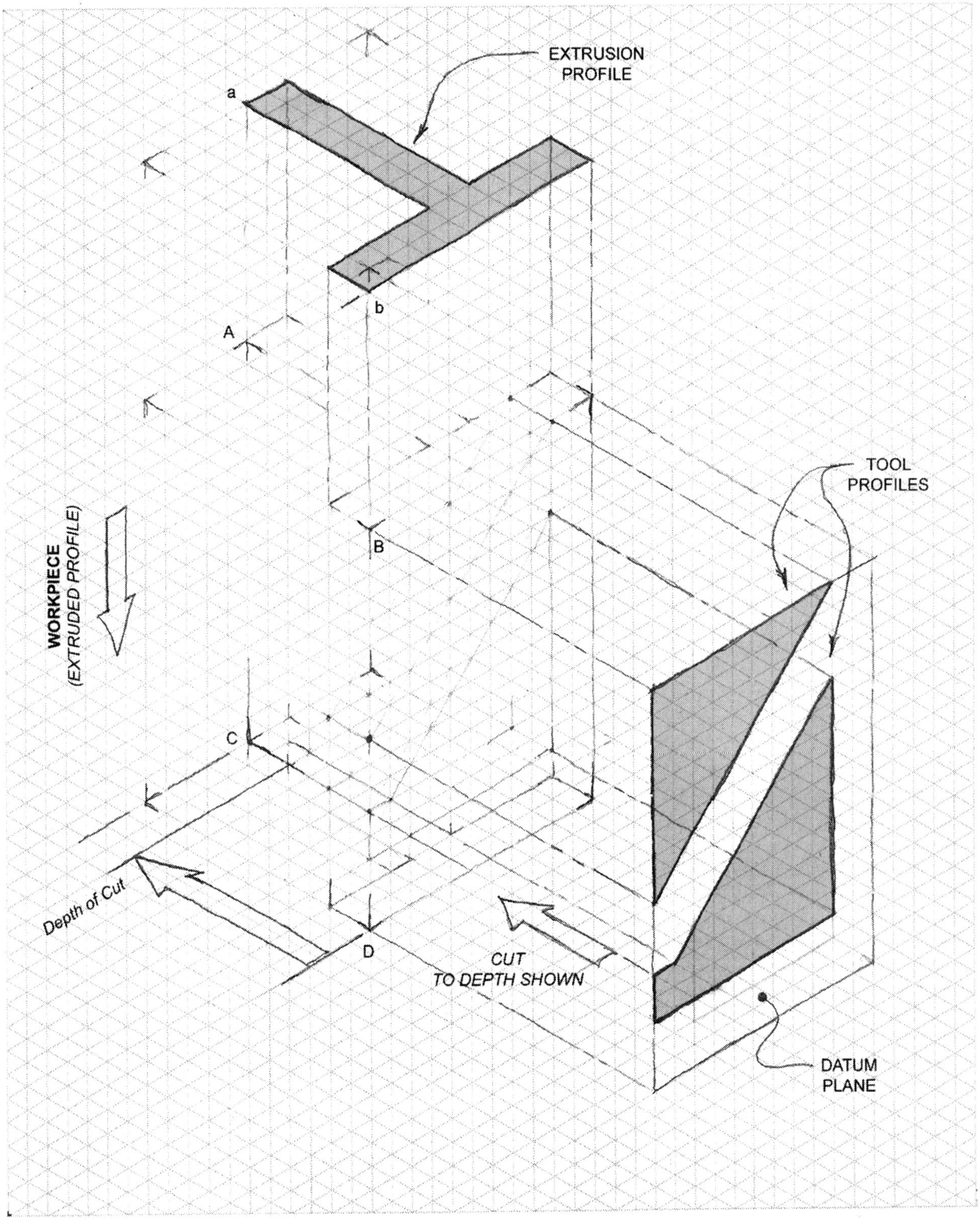

Problem 10.4

Drawn By:		FREEHAND SKETCHING	
	Scale	TITLE FABRICATION	
Date	Sheet	Drawing Number A	10.4

PATTERN FOLDING & LAYOUT: Given orthographic and isometric pictorial sketches of "folded" thin walled objects, sketch the layout for the pattern of the object in the space provided. Label each surface on your pattern to match the labeled surfaces and add tabs for seams as required. Plan the pattern to fit in the space allowed in the gridded area. If assigned, cut out the pattern, fold it together, and attach tabs to verify that it is correct.

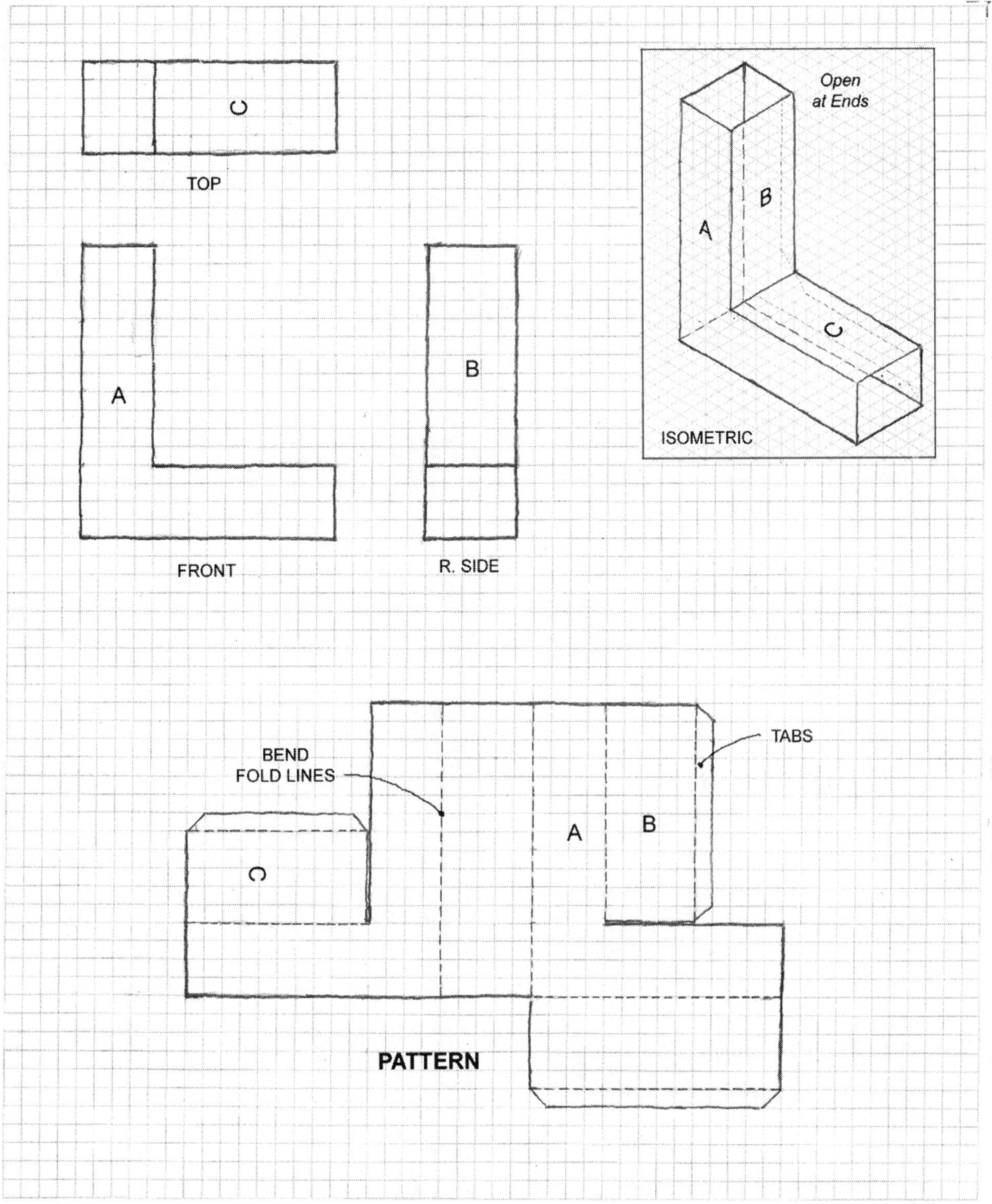

Problem 10.5 ELBOW	Drawn By:		FREEHAND SKETCHING	
EXAMPLE		Scale	TITLE PATTERN LAYOUT	
	Date	Sheet	Drawing Number A	10.5

PATTERN FOLDING & LAYOUT: Given orthographic and isometric pictorial sketches of "folded" thin walled objects, sketch the layout for the pattern of the object in the space provided. Label each surface on your pattern to match the labeled surfaces and add tabs for seams as required. Plan the pattern to fit in the space allowed in the gridded area. If assigned, cut out the pattern, fold it together, and attach tabs to verify that it is correct.

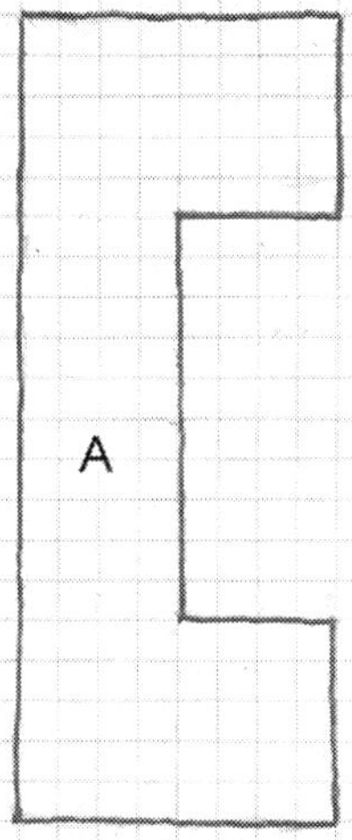

FRONT

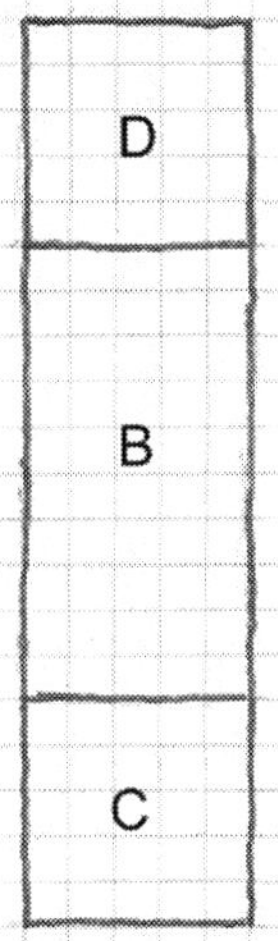

R. SIDE

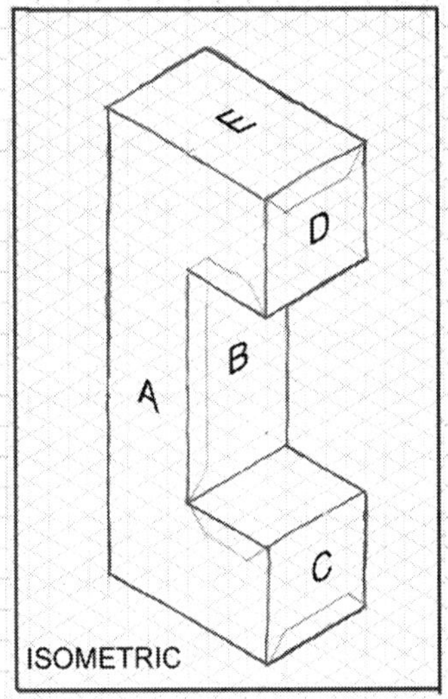

ISOMETRIC

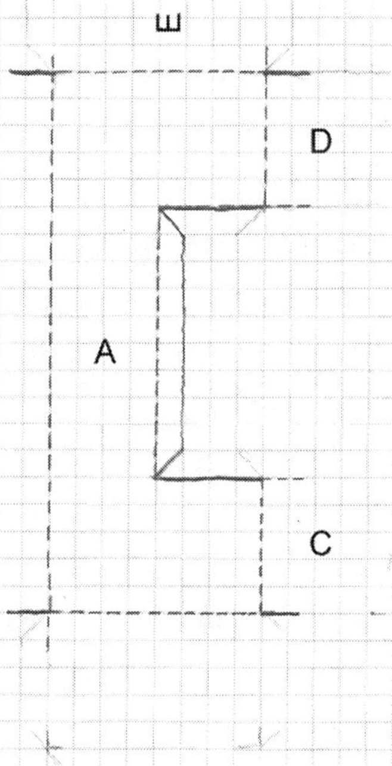

PATTERN

Problem 10.6.

Drawn By:		FREEHAND SKETCHING	
	Scale	TITLE PATTERN LAYOUT	
Date	Sheet	Drawing Number A	10.6

PATTERN FOLDING & LAYOUT: Given orthographic and isometric pictorial sketches of "folded" thin walled objects, sketch the layout for the pattern of the object in the space provided. Label each surface on your pattern to match the labeled surfaces and add tabs for seams as required. Plan the pattern to fit in the space allowed in the gridded area. If assigned, cut out the pattern, fold it together, and attach tabs to verify that it is correct.

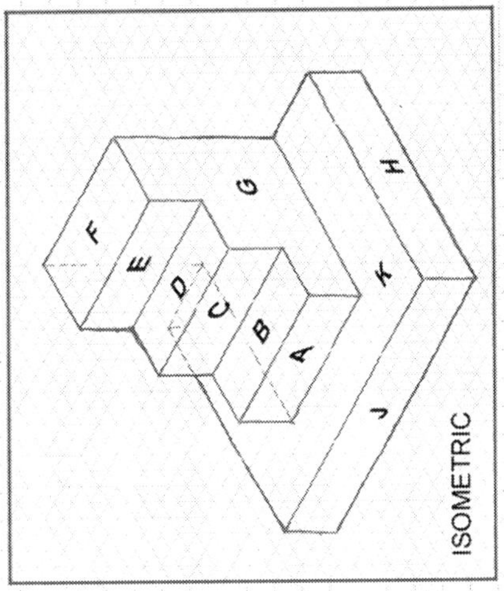

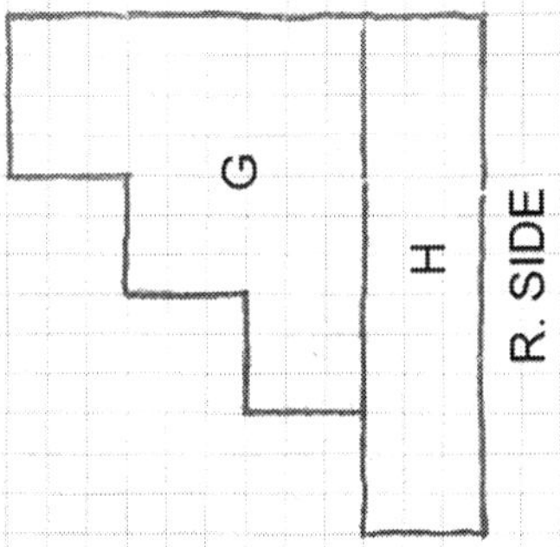

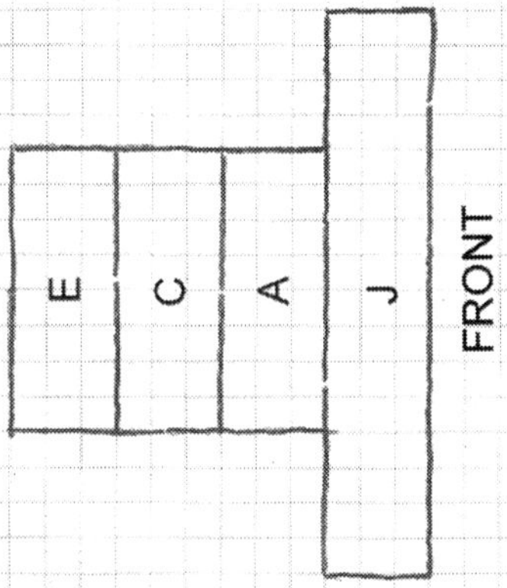

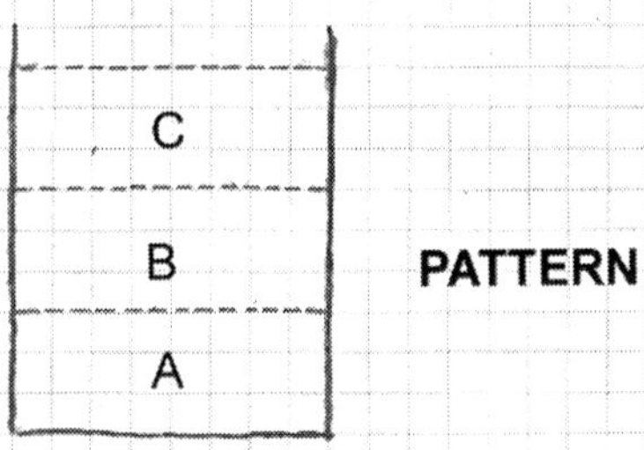

Problem 10.7.

Drawn By:		FREEHAND SKETCHING	
	Scale	TITLE PATTERN LAYOUT	
Date	Sheet	Drawing Number A	10.7

PATTERN FOLDING & LAYOUT: Given orthographic and isometric pictorial sketches of "folded" thin walled objects, Sketch the layout for the pattern of the object in the space provided. Label each surface on your pattern to match the labeled surfaces and add tabs for seams as required. Plan the pattern to fit in the space allowed in the gridded area. If assigned, cut out the pattern, fold it together, and attach tabs to verify that it is correct.

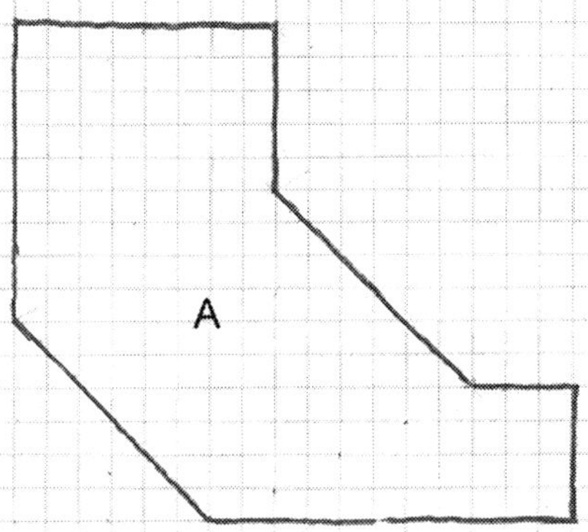

FRONT

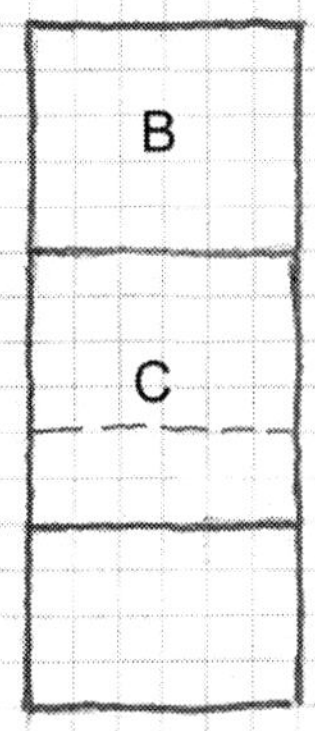

R. SIDE

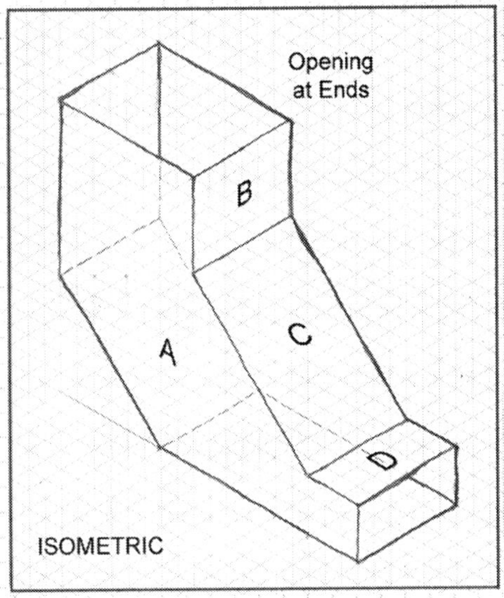

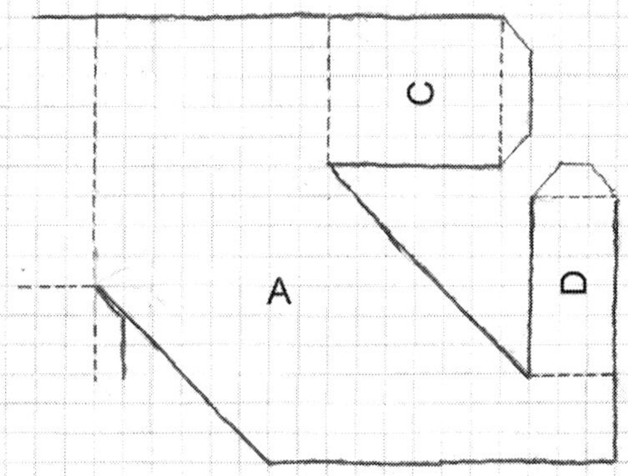

PATTERN

Problem 10.8.

Drawn By:		FREEHAND SKETCHING	
	Scale	TITLE PATTERN LAYOUT	
Date	Sheet	Drawing Number A	10.8

Drawn By:		FREEHAND SKETCHING	
	Scale	TITLE	
Date	Sheet	Drawing Number A	

Drawn By:		FREEHAND SKETCHING	
	Scale	TITLE	
Date	Sheet	Drawing Number A	

Drawn By:		FREEHAND SKETCHING	
	Scale	TITLE	
Date	Sheet	Drawing Number A	

Drawn By:		FREEHAND SKETCHING	
	Scale	TITLE	
Date	Sheet	Drawing Number A	

Drawn By:		FREEHAND SKETCHING	
	Scale	TITLE	
Date	Sheet	Drawing Number A	

Drawn By:		FREEHAND SKETCHING	
	Scale	TITLE	
Date	Sheet	Drawing Number A	

Drawn By:		FREEHAND SKETCHING	
	Scale	TITLE ORTHOGRAPHIC GRID SHEET	
Date	Sheet	Drawing Number A	

Drawn By:		FREEHAND SKETCHING	
	Scale	TITLE ORTHOGRAPHIC GRID SHEET	
Date	Sheet	Drawing Number A	

Drawn By:		FREEHAND SKETCHING	
	Scale	TITLE ORTHOGRAPHIC GRID SHEET	
Date	Sheet	Drawing Number A	

Drawn By:		FREEHAND SKETCHING	
	Scale	TITLE ORTHOGRAPHIC GRID SHEET	
Date	Sheet	Drawing Number A	

Drawn By:		FREEHAND SKETCHING	
	Scale	TITLE ORTHOGRAPHIC GRID SHEET	
Date	Sheet	Drawing Number A	

Drawn By:		FREEHAND SKETCHING	
	Scale	TITLE ORTHOGRAPHIC GRID SHEET	
Date	Sheet	Drawing Number A	

Drawn By:		FREEHAND SKETCHING	
	Scale	TITLE ISOMETRIC GRID SHEET	
Date	Sheet	Drawing Number A	

Drawn By:		FREEHAND SKETCHING	
	Scale	TITLE ISOMETRIC GRID SHEET	
Date	Sheet	Drawing Number A	

Drawn By:		FREEHAND SKETCHING	
	Scale	TITLE ISOMETRIC GRID SHEET	
Date	Sheet	Drawing Number A	

Drawn By:		FREEHAND SKETCHING	
	Scale	TITLE ISOMETRIC GRID SHEET	
Date	Sheet	Drawing Number A	

Drawn By:		FREEHAND SKETCHING	
	Scale	TITLE ISOMETRIC GRID SHEET	
Date	Sheet	Drawing Number A	

Drawn By:		FREEHAND SKETCHING	
	Scale	TITLE ISOMETRIC GRID SHEET	
Date	Sheet	Drawing Number A	